America Transformed

Engineering and Technology in the Nineteenth Century

Selections from the
Historic American Engineering Record,
National Park Service

Dean Herrin

ASCE PRESS
American Society of Civil Engineers
1801 Alexander Bell Drive / Reston, Virginia 20191-4400

ABSTRACT: America was transformed in a variety of ways in the nineteenth century by engineering and technology. From a fledgling, agrarian, and largely unexplored nation clinging to the Atlantic seaboard in 1800, the United States by 1900 was the world's leading industrial nation, manufacturing a third of the entire world's industrial output, and on its way to becoming a more urbanized and mechanized country. Since 1969, the Historic American Engineering Record of the National Park Service has documented, through photographs, drawings, and histories, thousands of significant American engineering and industrial structures and sites. This book presents a sampling of that documentation to show the scope and variety of sites, structures, mechanisms, and complexes that were components of America's industrial transformation in the nineteenth century.

LIBRARY OF CONGRESS CATALOGING-IN-PUBLICATION DATA

Herrin, Dean A., 1958–
America transformed : engineering and technology in the nineteenth century / Dean Herrin.
p. cm.
"Selections from the Historic American Engineering Record, National Park service."
Includes bibliographical references and index.
ISBN 0-7844-0529-8
1. Engineering—United States—History—19th century. 2. Technology—United States—History—19th century. I. United States. National Park Service. National Architectural and Engineering Records Division. II. Title.

TA23 .H47 2002
609.73'09'034—dc21 2002027963

[FRONT COVER] *Background photograph:* Carding machine, c. 1865, Watkins Woolen Mill, Lawson, Missouri (photograph by Jet Lowe). *Inset images, left to right:* New Croton Dam and Reservoir, New York (photograph by Jack Boucher); Harmony Manufacturing Company, Cohoes, New York (photocopy by Jack Boucher; original from Evan Leigh, *Modern Cotton Spinning*); Steam engine, St. John, Virgin Islands (photograph by Jet Lowe); and Brooklyn Bridge, New York City (photograph by Jet Lowe).

[BACK COVER] *Inset images, left to right:* Smithfield Street Bridge, Pittsburgh, Pennsylvania (photograph by Jack Boucher); Block Island Southeast Light, Rhode Island (drawing by Lee Ann Jackson and Isabel C. Yang); Turbine governor, Folsom Powerhouse, California (photograph by Brian Grogan); and work crew, 1895, Ashtabula, Ohio (photograph courtesy of Mrs. L. J. Fargo, Lexington, OH).

Contents

This book is dedicated to the memory of

Neal FitzSimons,

long a bridge between the American Society of Civil Engineers and the Historic American Buildings Survey / Historic American Engineering Record.

Foreword

This book is intended to reveal the extremely rich and varied treasures found in the Historic American Engineering Record (HAER). Though little known by the American public, it is one of the most comprehensive graphic and textual collections elucidating America's engineering, industrial, and technological achievements. This book will help alleviate this lack of familiarity. The book is authored by Dean Herrin, who served as HAER staff historian for ten years. *America Transformed* focuses on HAER's documentation of historic sites and structures related to nineteenth-century American engineering and technology. The book's publication coincides with the sesquicentennial celebration of one of America's oldest professional organizations. Established 150 years ago, the American Society of Civil Engineers was created to remind people that while this nation was founded on principles of life, liberty, and the pursuit of happiness, that vision could not be fulfilled without the contribution of engineers.

It has been said that the past is prologue. HAER was established thirty-three years ago to preserve the evidence of the engineer's prowess along with the nation's other technological achievements. Though HAER is not as old as ASCE, its establishment at the highest level of government and its partnership with other esteemed organizations gives legitimacy and authority to the function of compiling an archive. We only can aspire to continue these achievements by knowing what was accomplished in the past. The program is vested within the National Park Service, along with the other conservation and preservation programs of the United States, with support from the American Society of Civil Engineers, its major constituent group, and many of the other engineering organizations. The collection is curated and made available to the American public by the Library of Congress. That archive is represented by nearly 3,500 sheets of measured and interpretive drawings, 72,000 large-format photographs, 61,000 data pages, and 1,000 color transparencies, comprising information on nearly 7,000 sites, structures, and objects. This rich archival treasure serves as the cornerstone of this book.

The author has selected not just the obvious icons of civil engineering such as the Brooklyn Bridge, the Lowell textile mills, and Erie Canal, but the not so obvious industrial and technological feats such as the textile, iron and steel, and wind mills; public works such as water treatment, irrigation, and hydroelectric generation; and, the canal, railroad, and highway infrastructure that knit us together. We take much for granted when flipping a light switch or turning the faucet. This book begins to reveal what's behind those seemingly everyday conveniences.

A remarkable thing about the HAER collection is that it is based on the painstaking measurement, research, and documentation of actual works. This means that many of the sites, structures, and objects depicted are still standing. Thus the mission of HAER is not just preservation through documentation, but also to promote and encourage the commemoration and preservation of selected monuments and artifacts of the engineer's, industrialist's, and laborer's art. A goal of this book is to encourage the public to notice and better understand the meaning and value of these types of resources, and, when appropriate, to advocate the preservation of selected examples.

HAER also strives to introduce students to this great engineering and technological legacy. One of the largest student internship programs of the federal government has been implemented by HAER and the National Park Service. Over three decades, three thousand student engineers, architects, and historians of technology have served on HAER teams creating the histories, drawings, and photographs that comprise the HAER collection. Anyone having served on a HAER team for twelve weeks will never look at a bridge, dam, canal, factory, or power plant in similar light.

When Librarian of Congress, Dr. James Billington, commissioned the Kellogg Foundation to determine which of its vast collections should be first tapped for the National Digital Library, the HAER collection was singled out because of its national coverage, its quality, consistency, and easily reproduced format, and the fact that it could be used by primary and secondary school students K-12 for local history and science projects. It also deals with engineering and technology, one of the career choice areas that has declined among young people. By producing this and subsequent volumes with ASCE, we at HAER hope to reveal the remarkable legacy of our engineers, laborers, and technologists. While it is not possible to preserve all remaining engineering and industrial artifacts, there is at least comfort in knowing that a permanent record will be available to future generations through the internet and at the Library of Congress.

ERIC DELONY
Chief, Historic American Engineering Record
National Park Service
Spring 2002

Acknowledgments

Library research can have its slow moments, but sitting in the reading room of the Prints and Photographs Division of the Library of Congress, poring over more than 30 years of work by the Historic American Engineering Record (HAER), was a task I immensely enjoyed. Even after a decade as a historian for HAER, I was only dimly aware of the magnitude and richness of the collection. Sampling quickly gave way to a project-by-project examination, even as librarians looked a bit concerned as I wheeled cart after cart to my table. My thanks first go to those librarians in the Prints and Photographs Division, both for their assistance and forbearance during my research and, mostly, for their dedication in maintaining the Historic American Buildings Survey / Historic American Engineering Record (HABS/HAER) collection and in making it available to the public.

The collection exists, of course, because HAER has fielded team after team, year after year, to document America's engineering and industrial legacy. After reviewing their work, I appreciate more than ever the work of the hundreds of project team members who over the years have added their talents and unique contributions to this endeavor. I especially would like to thank Eric DeLony, Chief of HAER, for his continued enthusiasm and commitment in leading the organization. Eric was HAER's first permanent employee, hired in 1971, and he has been an unwavering advocate for the role HAER plays in both the documentation and preservation of historic technological sites and in the training of young professionals in the field. Robert Kapsch was the first to suggest a book on HAER's documentation of significant nineteenth century sites, and Blaine Cliver, Chief of HABS/HAER, has supported the project from the beginning. I thank both Eric and Blaine for the time away from normal duties to complete this book.

Even though I have worn their collective patience thin, current and former staff members of HABS/HAER have helped me tremendously. I would like to thank Tom Behrens, Brian Cary, Brian Cleven, Todd Croteau, Tim Davis, Gray Fitzsimons, Catherine Lavoie, Dana Lockett, Christopher Marston, Rich O'Connor, and Justin Spivey for reviewing drafts, answering questions, finding files, listening to a writer's complaints, and graciously forgiving me when I hogged the office copies of project files and forgot to leave a sign-out card. The HABS/HAER staff photographers, Jet Lowe and Jack Boucher, are wonderful colleagues and gave me very helpful advice in the selection of photographs. James Rosenthal did a superb job of printing the photographs. I would especially like to thank Lisa Davidson and Martin Perschler for all of their assistance in preparing this book.

The American Society of Civil Engineers (ASCE) was one of the founders of HAER in 1969, and the organization's support has made this book possible. I owe a debt to Jack Bruggeman, Alyson Hackett, and Suzanne Coladonato for their guidance. I would also like to thank the National Park Foundation, for its early support in sponsoring the project. I also appreciate reviews and advice from Steve Lubar, Robert Vogel, Helena Wright, Neal FitzSimons, and Terry Reynolds.

Finally, I most want to thank Sarah Heald and Emma Herrin, loving advisors.

Dean Herrin
National Park Service Coordinator
Catoctin Center for Regional Studies
Frederick Community College
Frederick, Maryland

America Transformed

INTRODUCTION

The Hedgehog, the Fox, and the HAER

The historian Isaiah Berlin began his study of Leo Tolstoy's conception of history by quoting an ancient Greek proverb: "The fox knows many things, but the hedgehog knows one big thing."[1] These cryptic words seem to imply that despite all the fox's knowledge and cunning, he can still be defeated by the hedgehog's one big defense—his quills. Berlin, however, suggests that the parable also can be used to distinguish between two opposing views of how people interpret the world. One group, represented by the fox, pursues many different ends, and interprets events as diffuse, fragmentary, and unrelated; the other group, symbolized by the hedgehog, categorizes everything into a single coherent and unified vision. Berlin's conclusion about Tolstoy is that the great writer was a fox who longed to be a hedgehog.

To borrow Berlin's dichotomy, the story of American engineering and technology in the nineteenth century is that of a fox that was transformed into a hedgehog. In a century of immense changes, perhaps the greatest transformation for Americans in the nineteenth century was technological. From a world of isolation, self sufficiency, and local, small-scale production, by 1900 Americans had passed into a technological cornucopia that was inextricably linked with a new world of interdependence, abundance, system, and order. By century's end Americans had reached "the promised land of the technological world, the world as artifact."[2] The United States was far from completely unified and centralized at the end of the nineteenth century, but it was so much more so than at the beginning of the century as to constitute a break with the previous order. As the underpinning for the larger process of industrialization that had been occurring in America since the late eighteenth century, technology itself was transformed from a fragmentary and diffuse presence into an increasingly pervasive, rationalized, and centralized system as the century progressed. According to the historian of technology Ruth Schwartz Cowan:

> Early in the nineteenth century the process of industrialization had appeared ... as a rather discrete undertaking: a spinning factory in a neighboring town, a merchant miller up the river, a railroad station a few miles distant. By the end of the century, virtually all Americans must have been aware that it had become something vastly different: a systematic undertaking that had created interlocking physical and social networks in which all Americans—rich

From a world of isolation, self sufficiency, and local, small-scale production, by 1900 Americans had passed into a technological cornucopia that was inextricably linked with a new world of interdependence, abundance, system, and order.

From a fledgling, agrarian, and largely unexplored nation clinging to the Atlantic seaboard in 1800, the United States was the world's leading industrial nation by 1900, manufacturing a third of the world's industrial output.

or poor, young or old, urban or rural—were increasingly enmeshed.[3]

One of the dominant themes of nineteenth-century American history, in fact, is this creation of "interlocking physical and social networks." In *The Incorporation of America*, Alan Trachtenberg argues that there emerged in America in the three decades after the Civil War a "changed, more tightly structured society with new hierarchies of control, and also changed conceptions of that society, of America itself."[4] Yet technology was not autonomous, an unseen force acting on unsuspecting Americans. In a complex interaction, technology both reflected and shaped cultural values of the era. Walter Licht, in a recent synthesis of American industrialization in the nineteenth century, has stressed that a convergence of efforts by various groups, many with opposing agendas, led to the creation of a "new" United States, one based on "greater security and a more administered economy and polity."[5] Robert Wiebe, in *The Search for Order*, claims that the transformation to a more ordered and structured society occurred because Americans were searching for "continuity and predictability in a world of endless change."[6]

The "world of endless change" was due in large measure to technology. Whether one looks at American industrialization in the nineteenth century in general, or at various components, such as urbanization, manufacturing, transportation, or the imperatives of a new capitalist economy, technology, as the embodiment of a dominant cultural ethos, can usually be found as the cause of change, or a means by which change was effected, or both. Statistics reveal that the technological transformation was enormous. From a fledgling, agrarian, and largely unexplored nation clinging to the Atlantic seaboard in 1800, the United States was the world's leading industrial nation by 1900, manufacturing a third of the world's industrial output.[7] Americans in that year produced more than a third of the world's supply of steel, more than England and Germany combined.[8] Railroad mileage in the United States extended over 250,000 miles by 1900, more than for all of Europe combined.[9] As new energy sources freed industry from the water power of the countryside, and as the ever-broadening transportation network created geographic nexuses, Americans became more urbanized as well. By 1900, almost 40% of the nation's total population (more than 30 million Americans) lived in cities of 2,500 people or more, double the percentage from 1860.[10]

A more evocative portrait of the effects of American industrialization can be found in an 1841 gazetteer, which describes a traveler's experience upon first visiting Pittsburgh:

> A dense cloud of darkness and smoke ... visible for some distance before he reaches it, hides the city from his eyes until he is in its midst; .... As he enters the manufacturing region, the hissing of steam, the clanking of chains, the jarring and grinding of wheels and other machinery, and the glow of melted glass and iron, and burning coal beneath, burst upon his eyes and ears in concentrated force. If he visits the warehouses, he finds glass, cotton yarns, iron, nails, castings, and machinery, occupying a prominent place. He discovers the whole city under the influence of steam and smoke.[11]

It is little wonder, then, that motion and change became synonymous with the American people in the nineteenth century. As early as 1830, Alexis de Tocqueville took notice of "everyone in motion" in a "universal tumult." Americans, he concluded, live "in a state of incessant change of place, feelings, and fortunes."[12] A less prosaic Indiana farmer remarked in the late 1860s, "Everybody and everything's goin' places."[13] By the late 1880s, the American author Hamlin Garland sounded a more alarming note: "Nothing is stable, nothing absolute, all changes, all is relative."[14]

By the late nineteenth century, Americans were unsettled yet optimistic, and both attitudes were largely shaped by technology. Constant change produced an unsettled feeling and a desire for stability and order, yet, even though technology was one of the root causes of most of the changes taking place in society, many Americans maintained an unshakable faith in technology. There were certainly those, like Henry Thoreau and Henry Adams, who were skeptical at best of the increasingly pervasive influence of technology on American society, and the economic crises of the latter part of the century coupled with laborers' and farmers' protests brought into question the direction in which the country was heading. Even as early as 1842, Washington Irving wanted to "blow up all the cotton mills ...

and make picturesque ruins out of them."[15] The enormous environmental and human costs of nineteenth-century technology were slow to register on most Americans, however. "Thoreau spoke for himself, but not for Pittsburgh," explains one historian.[16]

The people of Pittsburgh, in fact, and Boston, and New York, and most everywhere else in America, were hopeful and many were even enthusiastic about the emerging technology of the country. Historians have described the "exhilaration," the "technological enthusiasm," the "thrill of the technological transformation," and the almost religious fervor of nineteenth-century Americans captivated by engineering and technological developments.[17] Some engineering achievements, such as the Erie Canal and the Brooklyn Bridge, even achieved the status of "sublime," according to David Nye, eliciting in observers feelings of awe and wonder (and even terror at times).[18] Walt Whitman was positively giddy in his endorsement of modern technological achievements, "Singing the great achievements of the present, Singing the strong light works of engineers, / Our modern wonders."[19] Even Ralph Waldo Emerson, not altogether taken with technology, was inspired enough to express "pity for our fathers for dying before steam and galvinism."[20] Looking back on the period, the historian Perry Miller, in a marvelous rhetorical flourish, described Americans' enthusiasm for technology and how they "flung themselves into the technological torrent, how they shouted with glee in the midst of the cataract, and cried to each other as they went headlong down the chute that here was their destiny...."[21]

This faith, this fervor, in technology provided most Americans the vision for the ensuing "search for order" in the late nineteenth century. Americans became imbued and identified with the very attributes of technology. Thomas Hughes, for example, argues that "inventors, industrial scientists, engineers, and system builders have been the makers of modern America. The values of order, system, and control that they imbedded in machines, devices, processes, and systems have become the values of modern technological culture." This process, according to Hughes, was not a conspiratorial plot but simply a reflection of the dominant ethos of the time, of "all those, including [the inventors, industrial scientists, engineers, and system builders], who were uneasy in a seemingly chaotic and purposeless world and who searched for compensatory order."[22] Cecilia Tichi, likewise, claims that the dominant "gear-and-girder technology" of the late nineteenth and early twentieth century gave rise to a corresponding vision of "the material world as a dynamic, integrated assemblies of component parts," standardized and predictable. Americans looked to engineers, in fact, to provide an ordered and stable world.[23] "The machine had become the model not merely for industry, but for society as well," notes another historian.[24] The result, very broadly stated, was a more regulated, machine-like, and state-administered society by the end of World War I.[25]

We suffer now from technological sublime envy. In an age of invisible computer chips and all-too-complicated technology, Americans, or at least those who care about such things, miss the visual attractiveness and seemingly self-evident technological composition of nineteenth-century technology. The gear-and-girder technology of the last century was "visually compelling," and composed a world in which "[g]ears go round, bearings roll, pistons push, girders support."[26] A guidebook to the 1876 Philadelphia Centennial Exposition urged fairgoers to see the gigantic Corliss steam engine that powered all the machines in Machinery Hall:

> Poets see sublimity in the ocean, the mountains, the everlasting heavens; in the tragic elements of passion, madness, fate; *we* see sublimity in that great fly-wheel, those great walking-beams and cylinders, that crank-shaft, and those connecting rods and piston-rods,—in the magnificent totality of the great Corliss engine.[27]

One of those fairgoers described how the engine's "mighty walking-beams plunge their pistons downward, the enormous fly-wheel revolves with a hoarded power that makes all tremble, the hundred life-like details do their office with unerring intelligence."[28] No wonder *Scientific American* referred to the engine as "the poetry of mechanical motion."[29] Who today, when we have trouble determining if our machinery is even turned on, when we have difficulty simply *seeing* technology, would not want to experience the same?

This envy is part of the contradictory impulse in the second half of the twentieth century to study and pre-

In an age of invisible computer chips and all-too-complicated technology, Americans miss the visual attractiveness and seemingly self-evident technological composition of nineteenth-century technology.

Many have witnessed the rapid destruction of historic engineering and technological sites, and have become alarmed that we are obliterating and desecrating an important part of the American story, the roots of our modern built world.

serve American engineering and technological history. On the one hand, many have witnessed the rapid destruction of historic engineering and technological sites, and have become alarmed that we are obliterating and desecrating an important part of the American story, the roots of our modern built world. On the other hand, others (sometimes the same) want to study technology to better understand how and when we allowed ourselves to become a more regulated, machine-like, and state-administered society, beholden to technology. Just how did technology, they ask, become imbued with so many values, some contradictory? Is technology sublime, or rather the root of many of our societal problems? Since the late 1950s, there has been a growing interest in the engineering and technological history of the country, evidenced by the creation of such organizations as the Society for the History of Technology, the Society for Industrial Archeology, and the American Society of Civil Engineers' Committee on the History and Heritage of Civil Engineering; the opening of the Smithsonian Institution's Museum of History and Technology (now the National Museum of American History); the creation of graduate programs in the history of technology; the development of technological museums and sites such as that in Lowell, Massachusetts, administered by the National Park Service; and the plethora of books, articles, and journals devoted to the topic.

In the context of this burgeoning interest in America's engineering and technological history, the Historic American Engineering Record (HAER) was created in 1969 by a tripartite agreement among the National Park Service, the American Society of Civil Engineers (ASCE), and the Library of Congress, as a sister program to the Historic American Buildings Survey (HABS) of the National Park Service. Through documentation consisting of written histories, large-format photographs, and measured and interpretive drawings, HAER's mission is to record for posterity the enormous historic engineering and industrial legacy of the United States. As of 2002, HAER has documented 7,432 structures and sites through 73,120 photographs, 3,620 sheets of drawings, and 62,436 pages of written histories. All of this documentation is stored in the Prints and Photographs Division of the Library of Congress, copyright-free and available to the public. (The Library of Congress's "American Memory" website includes a searchable on-line catalog of the HABS/HAER collection, "Built in America: Historic American Buildings Survey/Historic American Engineering Record, 1933–Present." Go to <www.cr.nps.gov/habshaer>.)

HAER's photographers, architects, and historians have certainly been prolific, but why is their work significant? First, America's engineering and technological legacy is like an endangered species, constantly threatened with extinction. Many of the sites recorded in this book have already been demolished. Without the actual artifact, or documentation of that artifact, historians are deprived of a valuable research tool. Artifactual evidence—that information that is gleaned directly from examining an object—can help us better understand the technology of a particular era, and its associated values. Brooke Hindle eloquently argued that "physical objects are the expression of technology," much more so than they are in politics, science, religion, or other fields.[30] Technological artifacts and sites are not built with words, and words alone often can not describe how a technological artifact was built or functioned. Converting knowledge into practice is the essence of engineering, and artifacts are the evidence of that conversion.[31] "Artifacts ... show how something was done, not how someone said something was done," according to two leading practitioners of industrial archeology.[32]

In HAER's study of the Watkins Woolen Mill in Lawson, Missouri, for example, historian Larry Gross discovered several facts from the artifacts on the site that would have been difficult, if not impossible, to find in written documentary evidence.[33] Gross found, for example, a power loom which displayed an undocumented alternative to the standard power loom motion, as well as power looms that had been altered by workers at the Watkins Mill to improve the looms' performance. The site also revealed the "chaos" of working conditions in the mill, as the weave room contained seventeen looms of seven different types, and the awkward placement of large spinning jacks made the movement of workers difficult. The wide assortment of machinery in the mill further demonstrated the variety of options open to mid-nineteenth-century manufacturers, the tendency to mix old and new machinery, and evidence of technology that disappeared as improved machines came into existence. Wear patterns on the floor boards and walls scarred and pitted by errant flying shuttles also gave Gross a fuller understanding of the experiences of workers in the mill. In Gross' own words:

> To truly understand our technology and the working conditions and products it created, we must begin to examine the closest thing to primary evidence available, the artifacts which bring us closer than anything but a time machine to the history they record.[34]

Another primary contribution of HAER is its documentation of both the "hedgehog" and the "fox" qualities of nineteenth-century American engineering and technology. There are significant gaps in the HAER collection of particular sites, and even of particular industries and technologies, due to either the lack of survival of that particular site or industry, or to HAER's inattentiveness or inability to document certain sites. The HAER collection, though, supplemented by HABS's occasional documentation of engineering sites before the creation of HAER in 1969, is a remarkably comprehensive and broad record of America's technological past. Well-known examples of engineering achievement, like the Brooklyn Bridge and the Erie Canal, are included, but so are Dunlap's Creek Bridge, a small, obscure structure in Brownsville, Pennsylvania that was the first iron bridge built in America, and the Peavy-Haglin Grain Elevator in Minnesota, a sliver of a tower that introduced the concept of the reinforced-concrete grain elevator to this country. One can trace, for example, the evolution of the small New England textile mill of the first decade of the nineteenth century to the massive complexes that appeared later in the century; or discover the variety of ways in which early grist mills could be powered, from water to wind to tide; or follow the changes in technological materials from wood, stone, and brick to iron, steel, and concrete. The interconnectedness of late nineteenth-century technology is evident in the documentation of the iron mines, coal mines, and coke furnaces that fed the steel industry; the barges and rail lines that transported raw materials to the steel mills and carried finished products from them; the blast furnaces, open hearths, and forges of the mills themselves; the buildings and bridges and ships that used the steel fabricated in the mills; and the houses and neighborhoods in which the miners, steelworkers, and railwaymen lived. HAER strives to make these connections apparent, and to place sites and artifacts in their "hedgehog" context.

But one of the most important things that the HAER collection illuminates, by its site-by-site, artifact-by-artifact analysis, is the "foxness"—the diffuseness and variety—of American technology. The quest for the rational factory, the ordered industrial process, the regimented workforce did not happen all at once, and the images that follow also show a nineteenth-century America full of engineering variety, eclectic design and form, the occasional eccentric artifact, and even technologies treading down the wrong path. Reyner Banham has written that "far from being dominated by a reductive discipline of purely instrumental rationality, the high period of heavy industry was one of almost unlimited variability and richness."[35] Need a bridge in your town? Take your pick: suspension, truss, arch, cantilever, moveable, or covered, in wood, stone, iron, steel, or concrete? Would you like your water tower to look like a church, a castle, a tower, or a Doric column? There were, of course, engineering and environmental variables that figured into such decisions, but at times it depended only on how much money was available and the tastes and desires of those making the decisions. Technology does not determine itself. David Nye, for example, has argued that technical and economic factors alone cannot explain the evolution of the skyscraper. Equally important was the more ethereal "public enthusiasm for tall buildings."[36] The "teeming fecundity," in Banham's words, of nineteenth century engineering and technology, therefore, included grain elevators, machinery, and lean spare factories that later inspired Modernist architects and painters; structures and objects that are prime examples of the "picturesque eclectic," brimming with variety, movement, and irregularity; and other structures and sites that can only be called the unpicturesque eclectic, a motley assortment of utilitarian and nondescript buildings, towers, machines, and complexes.[37]

The quest for the rational factory, the ordered industrial process, the regimented workforce did not happen all at once; 19th-century America was full of engineering variety, eclectic design and form, the occasional eccentric artifact, and even technologies treading down the wrong path.

Technology is never neutral. The technological objects and sites that were built in the 19th century, the shape they took, how and where they were built, who built and controlled them, and who benefited most and least from them were all value-laden decisions.

The collections of the Historic American Engineering Record, and the images in this book, document this "teeming fecundity" of nineteenth-century technological America. This book is intended not as a comprehensive history of American industrialization, but as a visual sampler of America's nineteenth-century engineering and technology—to show the scope and variety of sites, structures, mechanisms, and complexes that were components of America's industrial transformation in the nineteenth century. Technology is never neutral. The technological objects and sites that were built in the nineteenth century, the shape they took, how and where they were built, who built and controlled them, and who benefited most and least from them were all value-laden decisions.[38] HAER's documentation itself is value-laden—what is documented, with how many photographs and drawings, from what angles, and with what historical narrative, for example, are questions inextricably intertwined with assumptions, values, and biases held by individual HAER staff members as well as by the institutional program itself.[39] Without glossing over these facts, however, or without ignoring the real environmental, political, and human costs and benefits of technology, perhaps one can look at the images in this book and remember Reyner Banham's homage to the water towers photographed by Bernd and Hilla Becher:

> Above all they manifest human ingenuity and almost inexhaustible inventiveness....We are looking here at a record—all too often ignored—of human creativity so multifarious, determined, and persistent that the only appropriate response is a decent humility.[40]

References

1. Isaiah Berlin, *The Hedgehog and the Fox* (New York: Simon and Schuster, n.d. [reprint of 1953 edition]), p. 1.

2. Thomas P. Hughes, *American Genesis: A Century of Invention and Technological Enthusiasm, 1870-1970* (New York: Viking, 1989), p. 1.

3. Ruth Schwartz Cowan, *A Social History of American Technology* (New York: Oxford, 1997), p. 171.

4. Alan Trachtenberg, *The Incorporation of America: Culture and Society in the Gilded Age* (New York: Hill and Wang, 1982), p. 4.

5. Walter Licht, *Industrializing America: The Nineteenth Century* (Baltimore: Johns Hopkins University Press, 1995) p. 168.

6. Robert H. Wiebe, *The Search for Order, 1877-1920* (New York: Hill and Wang, 1967), p. xiv.

7. Licht, *Industrializing America*, p. xiv.

8. Thomas A. Bailey and David M. Kennedy, *The American Pageant,* Vol. II, 8th edition (Lexington, MA: D.C. Heath, 1987), p. 515.

9. United States Department of Commerce, Bureau of the Census, *Historical Statistics of the United States, Colonial Times to 1970* (Washington: GPO, 1975), p. 731.

10. Carroll Pursell, *The Machine in America: A Social History of Technology* (Baltimore: Johns Hopkins University Press, 1995) p. 132.

11. Carroll Pursell, *Early Stationary Steam Engines in America* (Washington, D.C.: Smithsonian Institution Press, 1969) pp. 80–81.

12. Cecilia Tichi, *Shifting Gears: Technology, Literature and Culture in Modernist America* (Chapel Hill, NC: University of North Carolina, 1987), pp. 42–43.

13. Richard Hofstadter, William Miller, and Daniel Aaron, *The United States, Part 2* (Englewood Cliffs, NJ: Prentice-Hall, 1972), p. 482.

14. Garland quote from his 1886-87 unpublished essay, "The Evolution of American Thought," as reprinted in Tichi, *Shifting Gears*, p. 43.

15. Washington Irving quotation from Russell Blaine Nye, *Society and Culture in America, 1830-1860* (New York: Harper, 1974), p. 276.

16. Perry Miller, "The Responsibility of Mind in a Civilization of Machines," in Thomas P. Hughes, ed., *Changing Attitudes Toward American Technology* (New York: Harper and Row, 1975), p. 74.

17. Brooke Hindle, "The Exhilaration of Early American Technology," in Judith A. McGaw, ed., *Early American Technology: Making and Doing Things from the Colonial Era to 1850* (Chapel Hill, NC: University of North Carolina, 1994), pp. 40–67; Hughes, *American Genesis,* pp. 1–12; Miller, *Changing Attitudes Toward American Technology,* p. 67.

18. David E. Nye, *American Technological Sublime* (Cambridge, MA: MIT Press, 1994) pp. xi–xx.

19. Walt Whitman, "Passage to India," completed in 1870, as quoted in David S. Reynolds, *Walt Whitman's America* (New York: Alfred A. Knopf, 1995), p. 499.

20. Emerson quotation from Nye, *Society and Culture in America,* p. 265.

21. Perry Miller, "The Responsibility of Mind in a Civilization of Machines," p. 68.

22. Hughes, *American Genesis*, pp. 4–5.

23. Tichi, *Shifting Gears*, pp. 4, 105.

24. Richard D. Brown, "Modernization: A Victorian Climax," in Daniel Walker Howe, ed., *Victorian America* (Philadelphia: University of Pennsylvania Press, 1976), p. 42.

25. Licht, *Industrializing America*, p. 131.

26. Tichi, *Shifting Gears*, p. 5.

27. Richard Henry Stoddard, ed., *A Century After: Picturesque Glimpses of Philadelphia and Pennsylvania* (Philadelphia, 1876), as quoted in Kasson, *Civilizing the Machine*, p. 164.

28. William Dean Howells, "A Sennight of the Centennial," *Atlantic Monthly*, 38 (July 1876), as quoted in Kasson, *Civilizing the Machine: Technology and Republican Values in America, 1776–1990* (New York: Penguin, 1976) p. 165.

29. Nye, *American Technological Sublime*, p. 121.

30. Hindle, "The Exhilaration of Early American Technology," p. 48.

31. Elting Morison, *From Know-How to Nowhere: The Development of American Technology* (New York: Basic, 1974), p. 6.

32. Robert B. Gordon and Patrick M. Malone, *The Texture of Industry: An Archaeological View of the Industrialization of North America* (New York: Oxford University Press, 1994) p. 13.

33. Laurence F. Gross, U.S. Department of the Interior, Historic American Engineering Record (HAER), No. MO-1, "Watkins (Woolen) Mill," 1978, Prints and Photographs Division, Library of Congress, Washington, D.C.

34. Gross, "The Importance of Research Outside the Library: Watkins Mill, A Case Study," *IA, Journal of the Society for Industrial Archeology* 7 (1981):26.

35. Reyner Banham, "The Becher Vision," in Bernd and Hilla Becher, *Water Towers* (Cambridge, MA: MIT Press, 1988) p. 8.

36. Nye, *American Technological Sublime*, p. 89.

37. Banham, "The Becher Vision," p. 8; Carroll L.V. Meeks, *The Railroad Station: An Architectural History* (New Haven, CT: Yale University Press, 1956), pp. 1–2.

38. Gordon and Malone, *The Texture of Industry*, p. 11; Hughes, *American Genesis*, p. 4.

39. Richard Oestreicher, "Images of Ideology and the Ideology of Imagery," *Labour/Le Travail* 15 (Spring 1985), pp. 182–183.

40. Banham, "The Becher Vision," p. 7.

PROLOGUE

Discovering the New Technological World

On July 4, 1828, the venerable Marylander Charles Carroll, the last living signer of the Declaration of Independence, inaugurated in Baltimore the building of the Baltimore and Ohio Railroad, declaring, "I consider this among the most important acts of my life, second only to that of signing the Declaration of Independence, if, indeed, second to that."[1] Fifty years later, and somewhat less solemnly, the fictional character "Josiah Allen's wife" paid a similar homage to engineering and technology when she described seeing the gigantic Corliss steam engine at the Centennial Exposition in Philadelphia: "Why, if there hadn't been another thing in the hull buildin', that great 'Careless Enjun' alone, was enough to run anybody's idees up into majestic heights and run'em round and round into lofty circles and spears of thought, they hadn't never thought of runnin' into before."[2]

Many Americans, like "Josiah Allen's wife," discovered something about themselves in the nineteenth century: they were enthusiastic, even exhilarated at times, about the new technological world they inhabited. Technology may have represented a "machine in the garden," but for many, it was a welcome machine, for the garden seemed to need some tilling. For these people, "invention, technological change, and industrialization were exciting developments that appeared to have something to do with making the world more interesting, with opening more options, and with improving the quality of life," according to historians Brooke Hindle and Steven Lubar.[3] There were exceptions to this enthusiasm, of course, particularly among many who actually labored in the new factories, mills, and mines generated by technology, and among intellectuals such as Henry Adams who were skeptical and wary of the precise role of engineers and technology in American society. Still, for most Americans, technology in the nineteenth century was of intense interest and constant wonder. By century's end, many Americans were convinced that technology "could bring order out of chaos, provide boundless energy, support business enterprise, and win wars."[4]

Even before the end of the century, technology in nineteenth-century America became intricately identified with ideas of liberty, equality, and national progress.[5] The novelist William Dean Howells preferred the Corliss engine and other technological exhibits at the Philadelphia Centennial Exposition to the exhibits of painting and sculpture, exclaiming, "It is ... in these things of iron and steel that the national genius most freely speaks."[6] Great technological achievements such as the Erie Canal, the cotton mills of Lowell, and the Brooklyn Bridge became icons of America's greatness and, at times, even achieved the level of the technological "sublime," imbued with transcendent significance.[7] There was a religious fervor to this conception of technology, as when Walt Whitman spoke of "the great cathedral sacred industry."[8] Words, in fact, often seemed inadequate to describe the emotions Americans felt as they surveyed the new technological world. Contemplation alone seemed the requisite response. On first seeing a new power loom, one of Lowell's founders, in a "state of

Brooklyn Bridge, constructed between 1869 and 1883, spanning the East River between Brooklyn and Manhattan, New York City, New York.

The Brooklyn Bridge became an enduring symbol of American engineering achievement in the nineteenth century. Designed by John Roebling (1806-1869) and completed by his son Washington (1837-1926), the bridge was the longest suspension bridge in the world when completed.

admiration and satisfaction," sat "by the hour, watching the beautiful movement of this new and wonderful machine...;" after the completion of the Eads Bridge in St. Louis, Walt Whitman said he "haunted the river every night... where I could get a look at the bridge by moonlight... I never tire of it... it is so fascinating, dreamy;" and yet another awestruck observer of the Corliss engine in Philadelphia claimed that the machine was "the highest embodiment of man's power, and I could stand by the hour watching it."[9]

A "technological torrent" washed over Americans in the nineteenth century, but several engineering and industrial achievements stood above the rest as icons of the new technological world. Accomplishments such as the Brooklyn Bridge and the Erie Canal were causes for national celebration, and evoked in most Americans feelings of pride, wonder, and anticipation of what new marvel might follow. Americans would come to discover darker attributes of their technological world, of the human and environmental costs of industrialization, but for many people in the nineteenth century, their admiration and hopes for technology echoed Nathaniel Hawthorne's comment after visiting an iron foundry in 1856, "What great things man has contrived, and is continually performing! What a noble brute he is!"[10]

[RIGHT] *Steam engine, c. mid-nineteenth century, St. John, Virgin Islands.*

New sources of power spurred American engineering and technology in the nineteenth century. Steam technology, pioneered in Great Britain, became the dominant power source of the century. This steam engine, most likely British, provided power for a sugarcane crusher on the Estate Adrian in the Virgin Islands. Steam was not widely adopted as a power source in the extensive Virgin Islands sugar industry, however, until after the emancipation of slaves throughout the Danish West Indies in 1848.[11]

[LEFT] *Erie Canal, Lockport Locks, 1859, Lockport, New York.*

The Erie Canal was one of America's greatest engineering achievements in the nineteenth century. Commenced on July 4, 1817, the canal stretched for 365 miles across New York, between Albany and Buffalo. The Erie Canal was the longest canal in the world when completed in 1825. The waterway opened up the West to commerce and emigration, provided an important training ground for a generation of American engineers, and ushered in a canal-building craze throughout the country. One of the most difficult engineering challenges faced by the canal builders was at Lockport, where the canal had to accommodate a 70-foot change in elevation. This was accomplished by a dramatic flight of five double locks, carved from rock, which allowed canal boats to travel simultaneously in either direction.

[FACING PAGE]
Clock tower, 1865, between Mills No. 1 and No. 2 of Boott Cotton Mills, Lowell, Massachusetts.

[RIGHT] *Line shaft, pulleys, and belts for the hoisting of power-canal gates, Pawtucket Gatehouse, 1847, Northern Canal, Lowell, Massachusetts.*

The textile mills of Lowell, Massachusetts, became famous as one of the first large industrial centers in America. The first mills opened in 1822, and by 1835, 22 mills were in operation, all operated by an extensive water-power system.[12]

[LEFT] *Mount Royal Station Trainshed, 1894-1896, Baltimore and Ohio Railroad, Baltimore, Maryland.*

Railroads radically changed Americans' conceptions of space and distance, and gave rise to new building forms. The Mount Royal Station trainshed, completed in 1896, was one of the last gable-roofed trainsheds in America. The roof was supported by steel trusses with arched lower chords; each truss was 71 feet in length. Gable-roofed trainsheds gave way gradually in the late nineteenth century to barrel-vaulted "balloon" sheds, but both eventually were replaced by simple platform canopies because of cost and the corrosive effects of exhaust gases from steam locomotives.[13]

Fairmount Waterworks, 1794-1822, Philadelphia, Pennsylvania.

The first major waterworks system in the country was built in Philadelphia, Pennsylvania, between 1794 and 1822, following the designs of Benjamin Latrobe (1764-1820) and Frederick Graff (1774-1847). In 1822, Graff designed these waterworks buildings in an elegant Greek Revival style, described by Charles Dickens in 1842 as, "no less ornamental than useful." By 1836, the system supplied almost 200,000 people with water, using six waterwheels.[14]

[LEFT] *Interior framing, Statue of Liberty, 1886, Liberty Island, New York.*

[BELOW] *Observation windows in tiara, Statue of Liberty, 1886, Liberty Island, New York.*

Americans in the nineteenth century marveled not only at American engineering, but also at examples of engineering from other countries. The Statue of Liberty, for example, a gift from France to the United States, was recognized as an ingenious statue of classical design and colossal proportion, supported by innovative engineering technology. The exterior form of the statue, composed of copper sheathing, was designed by the sculptor Frederic Bartholdi, and the intricate interior framework of wrought iron was conceived and fabricated by Gustave Eiffel.[15]

Thomas A. Edison Laboratories, 1887, West Orange, New Jersey.

[RIGHT] *Machine Shop of Building No. 5.*

[BELOW] *Chemistry Lab and Balance Room of Building No. 2.*

Thomas Edison personifies the American inventor of the nineteenth century, but he was not a lone inventor churning out inventions in his backyard. He developed, in fact, one of the first modern industrial research laboratories in Menlo Park, New Jersey, in the 1870s, followed by the establishment of this extensive laboratory in West Orange, New Jersey, in 1887. Along with a staff of about 60 researchers, Edison perfected in these labs the practical phonograph, the movie camera, the improved storage battery, and the fluoroscope, among many other inventions. Edison's West Orange facilities are now a National Historic Site administered by the National Park Service.

References

1. David E. Nye, *American Technological Sublime* (Cambridge, MA: MIT Press, 1994) p. 43.

2. M. Holley, *Josiah Allen's Wife as a P.A. and P.I.: Samantha at the Centennial.* Quoted in John F. Kasson, *Civilizing the Machine: Technology and Republican Values in America, 1776-1900* (New York: Penguin, 1976) p. 164.

3. Brooke Hindle and Steven Lubar, *Engines of Change: The American Industrial Revolution, 1790–1860* (Washington, D.C.: Smithsonian Institution Press, 1986) p. 93.

4. Thomas P. Hughes, "Introduction," *Changing Attitudes toward American Technology* (New York: Harper and Row, 1975) p. 9.

5. John F. Kasson, *Civilizing the Machine: Technology and Republican Values in America, 1776–1900.*

6. William Dean Howells, *A Hazard of New Fortunes* (1890), as quoted in Alan Trachtenberg, *The Incorporation of America: Culture and Society in the Gilded Age* (New York: Hill and Wang, 1982) p. 47.

7. Nye, *American Technological Sublime*, pp. xiii, 32.

8. Walt Whitman, "Song of the Exposition," (1871), as quoted in Nye, *American Technological Sublime*, p. xix.

9. Nathan Appleton, *Introduction of the Power Loom, and Origin of Lowell* (Lowell, 1858), as quoted in Kasson, *Civilizing the Machine*, p. 67.

10. Nathaniel Hawthorne, *English Notebooks*, ed. Randall Stewart (New York, 1941), as quoted in Kasson, *Civilizing the Machine*, p. 142; Walt Whitman, *Specimen Days* (1882), as quoted in Nye, *American Technological Sublime,* p. 83; "The Exhibition," *Nation,* 22(May 18, 1876), as quoted in Kasson, *Civilizing the Machine,* p. 164.

11. John Rumm, U.S. Department of the Interior, Historic American Engineering Record (HAER), No. VI-7, "The Development of the Sugar Cane Industry in the Virgin Islands," 1977, pp. 2–9, 12–15, Prints and Photographs Division, Library of Congress, Washington, D.C.

12. Hindle and Lubar, *Engines of Change*, p. 196.

13. Carl W. Condit, *American Building Art: The Nineteenth Century* (New York: Oxford University Press, 1960), pp. 206, 222.

14. Daniel L. Schodek, *Landmarks in American Civil Engineering* (Cambridge, MA: MIT Press, 1987), p. 202.

15. Condit. *American Building Art: The Nineteenth Century*, p. 46.

1

A Wealth of Resources

Even after 200 years of settlement, America had barely tapped its immense natural resources by the beginning of the nineteenth century. Land, water, forests, minerals—few other places on earth had such an abundant combination of life-sustaining and wealth-creating resources. Part of America's transformation in the nineteenth century was a changed conception of natural resources and a corresponding development of technologies to exploit that wealth.

The primary occupation of the American people in 1800 was agriculture. As industrialization proceeded in the nineteenth century, as more and more people moved to towns and cities and found work in factories, mines, and mills, the farm population declined. But agriculture remained a vital part of the American economy, and technological changes throughout the century made farming increasingly more productive. More farmland was opened in the last third of the century than in all of the nation's previous history.[1] New equipment such as the steel plow, the reaper, and the combine; improved scientific methods of farming; and larger concentrations of farmland dramatically increased production per farm. The amount of food that one farm worker could raise, for example, increased from enough to feed four people in 1820 to enough for seven people by 1900.[2]

An essential sector of the agricultural economy was the milling of grain. Grist mills had always been an important part of most American communities, and they continued to be so through most of the nineteenth century. Almost all were powered by water, although a few in coastal areas, such as on Long Island and Cape Cod, used the power of the wind or the tide. In 1840, there were close to 24,000 water-powered grist mills in the country, one for every 700 people.[3] In 1860, flour milling was still the country's leading industry in terms of the total monetary value of product.[4] Technologically, the most important improvement in milling occurred in the late eighteenth century, with Oliver Evans's invention of the automatic mill, which most millers eventually adopted because it reduced the cost of operation. With the adoption of Evans's system, milling remained essentially the same until the 1880s, when the roller mill process, which produced finer, whiter flour, was adopted from Hungary. As increasingly larger mill complexes were built to take advantage of economies of scale, primarily in Minneapolis, Minnesota, and other areas of the Midwest, smaller mills were unable to compete. Large merchant flour mills accounted for more than nine-tenths of America's flour by 1900.[5]

Exploiting the earth's resources in a different way, American mining industries flourished in the nineteenth century. The country's burgeoning technologies made unprecedented demands for coal, copper,

[FACING PAGE]

Hook Windmill, 1806, East Hampton, New York.

One of several windmills constructed by the Dominy family of Long Island, New York, this wind-powered gristmill operated for more than a century, until 1908.

Cotton gin, patented in 1844, Johnsonville, South Carolina.

iron, and other minerals, and mining companies responded with ever-larger operations, leading often to waste and inefficiency. Technological advances in ventilation, hoisting equipment, pumps, drills, mining methods, and electricity led to mines of unprecedented size, although complete mechanization and rationalization of the industries were difficult to obtain, and safety and environmental concerns were often neglected. Mining output was enormous, however, as coal surpassed wood in the mid-1880s as America's primary energy source. The great Mesabi iron range was opened in Minnesota in the 1880s and 1890s, and Michigan and a few western states yielded immense quantities of copper.[6]

Both a cause and effect of industrialization in the nineteenth century was an increasing concentration of people in cities. In 1874, the Unitarian minister David Wasson sadly observed that "the country is now but a suburb of the city." [7] Horace Greeley, editor of the *New York Tribune*, complained in the same era, "We cannot all live in cities, yet nearly all seem determined to do so...." [8] By century's end, most Americans in fact still lived in communities of fewer than 2,500 people, but the increasing dominance of the city was more pervasive than that population proportion would seem to indicate. Between 1830 and 1870, 18 cities in America grew to 100,000 or more inhabitants.[9] Transportation networks, labor supply, monetary institutions, the development of new power sources, and a myriad of other factors tended to concentrate factories (as well as other institutions) in urban areas. By 1900, nine-tenths of all manufacturing in the country occurred in cities.[10] In *The Age of Enterprise*, the historians Thomas Cochran and William Miller declared that "America, once a nation of farmers and small shopkeepers, agrarian in politics and provincial in spirit, had become by 1900 a land of big cities connected with one another by railroad, telegraph, and telephone and dominant in our national life."[11]

The growth of American cities in the nineteenth century was possible, in turn, because of technological advances in many areas, including urban transportation, water and gas supply, sanitation, building materials, and food preservation. Public water-supply systems, vital for public health as well as for fighting fires, were especially critical for the new urban centers. Innovative and acclaimed waterworks were built in Philadelphia, New York, Boston, and Washington, D.C. before the Civil War, but 128 other communities also had waterworks by 1860. By 1888, this number had mushroomed to 1,666.[12] Another significant part of the urban infrastructure was gas lighting. Baltimore was the first American city to use gas for lighting, in 1816, but it was not until after the Civil War that production, distribution, and metering issues were fully solved. Gas lighting eventually gave way to electric around the turn of the century, but the production and distribution networks established by gas companies provided the model for Thomas Edison and others in the new electric power industry.[13]

Harvesting the Land: Forest and Field

Up-and-Down Sawmill, mid-nineteenth century, near Chester Springs, Pennsylvania.

[ABOVE] *Detail of saw.*

[LEFT] *Exterior structure.*

The sawmill was a "particularly American form of mechanization," common in the United States but not in England and elsewhere.[14] America had an abundance of wooded land and a scarcity of artisan sawyers who opposed sawmills elsewhere. This sawmill in Chester County, Pennsylvania, was typical in that it used water power to move both the saw up and down and the carriage holding the log forward into the blade.

Rice Threshing Mill, c. 1840, Mansfield Plantation, near Georgetown, South Carolina.

[ABOVE] *Threshing mill with chimney stack, canal in foreground, and winnowing house in background.*

[RIGHT] *Steam-powered threshers.*

Rice plantations in South Carolina and Georgia adopted steam-driven threshing mills in the early nineteenth century to clean and separate rice grain from its husk. Intricate irrigation systems watered the rice fields of these plantations, and canals helped cart the rice from the fields to the mill.[15]

Browntown Cotton Gin, early nineteenth century, Johnsonville, South Carolina.

[LEFT] *Cotton gin building.*

[BELOW] *Gearing beneath gin building.*

The southern cotton industry was completely transformed in the early decades of the nineteenth century by Eli Whitney's invention of the cotton gin (short for "engine"), patented in 1793. Other cotton gins had been invented in different parts of the world, but none could easily process the short-staple cotton produced in the American South. Whitney's mechanical separation of seed from cotton fiber, along with improvements by others, made cotton production an enormously profitable enterprise, thereby ironically connecting this labor-saving device to the southern plantation system of slavery. Cotton gins could be powered by hand and, later, by steam, but most, such as this gin, were operated by horses or mules turning large wooden gears below the building housing the gin.

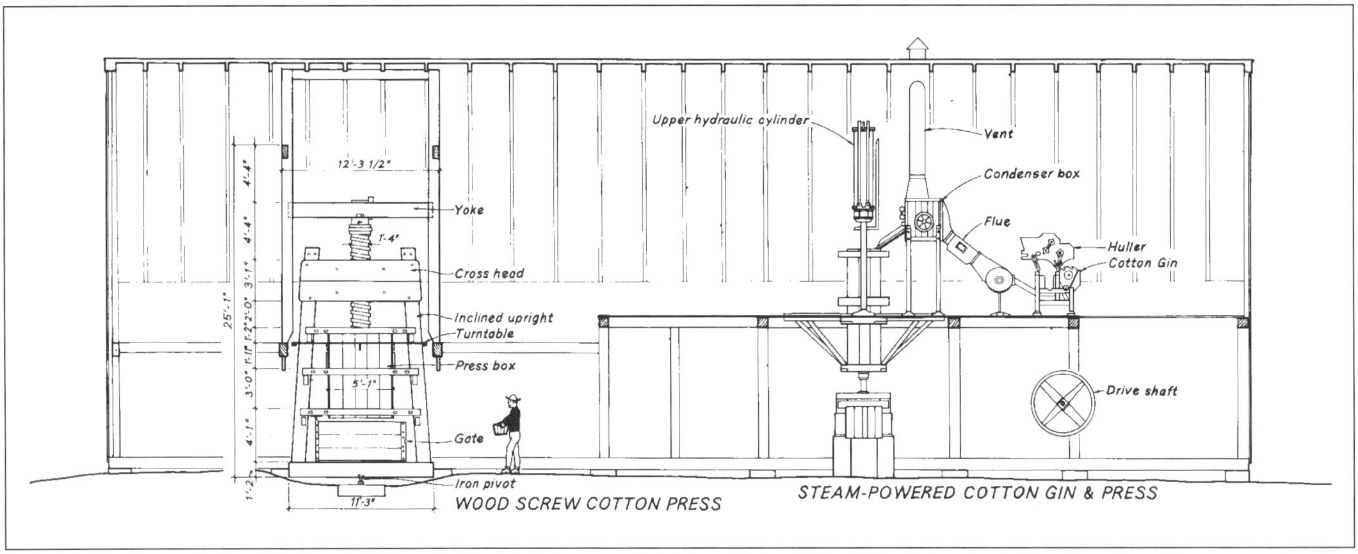

Magnolia Plantation Cotton Presses, Natchitoches, Louisiana.

[ABOVE] *Section of building containing wood-screw cotton press and steam-powered cotton press and gin.*

[RIGHT] *Wood-screw cotton press, c. 1830s.*

Early nineteenth century cotton shippers charged growers by volume instead of weight to transport bales of cotton, so to pack as much cotton into a bale as possible, growers turned to cotton presses. This wood-screw press in Louisiana, made of cypress timbers, was probably built in the 1830s. Most presses of this type were powered by either mules or horses. The press was part of Magnolia Plantation, a large antebellum cotton producer in Louisiana. In the years before the Civil War, the plantation produced more than 1,100 bales of cotton per year, each weighing approximately 450 pounds.

[RIGHT] *Cotton gins associated with the steam-powered cotton press.*

[BELOW] *Steam-powered cotton press, with ginning equipment in background.*

Agriculture, just as much as manufacturing and engineering sectors, underwent rapid mechanization and rationalization in the nineteenth century. Magnolia Plantation's antebellum wood-screw cotton press, for example, was supplemented in the latter years of the century by a steam-powered press that was part of a system to rationalize the ginning and baling process. The system included the steam press, cotton gins, and other equipment that automatically guided raw cotton from delivery wagons, through the gins, and into the press for forming into bales. Most of this equipment and system was the invention of Robert Munger, and according to HAER historian Richard O'Connor, his innovations "transformed ginning from a set of discrete steps to a continuous process."[16]

Contour planting, Arlington Heights orange grove, Riverside, California.

The citrus industry in southern California was made possible in the late nineteenth century by extensive irrigation systems and by the construction of highly engineered and controlled landscapes. In Riverside, California, the completion of the Gage Canal in 1890 marked the beginning of a vast citrus infrastructure that included groves and irrigation systems, along with packing houses, workers' camps, railroad lines, and machinery for fumigation control and frost prevention.

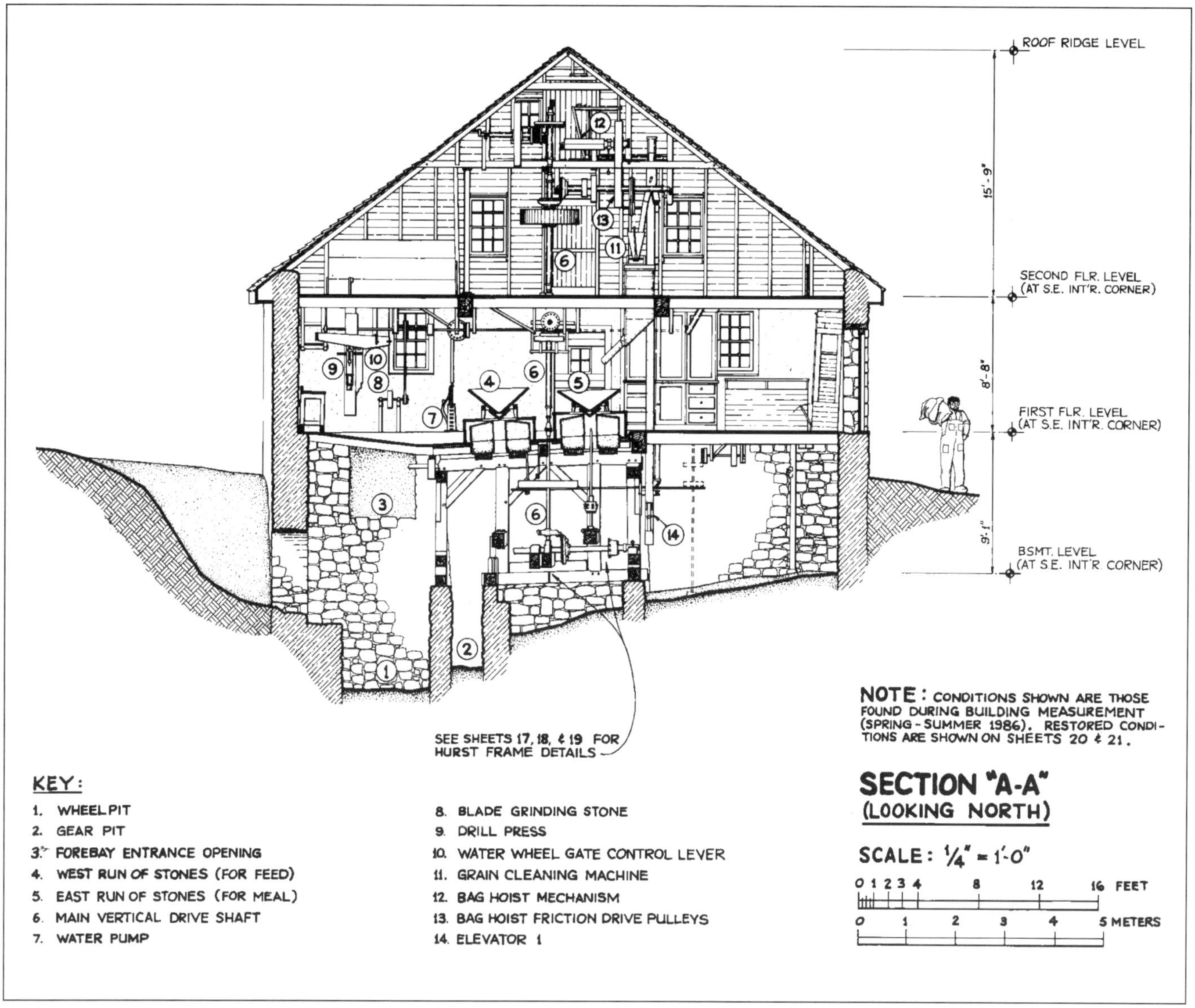

Section of Mount Pleasant Grist Mill, c. 1805, St. Peters, Pennsylvania.

This drawing represents the appearance of the partially dismantled Mount Pleasant Grist Mill in 1986 before restoration. When it was an operating mill, it primarily ground corn for both feed and meal. The key to milling was the flow of grain. To produce feed, grain was hoisted to the second floor, dropped into a hopper, which fed the grain into the feed stones [labeled 4 in the drawing], and then dropped into an elevator extending below the first floor and elevated back to the first floor for bagging. Grinding for meal was a slightly more complicated process. Grain was carried onto the first floor, dropped down a spout into an elevator [14], carried to the second floor to a grain cleaning machine [11], and dropped back down to the first floor via a hopper and spout to be ground in the meal stones [5]. After grinding, the meal was lifted by another elevator back to the second floor to be either bagged or further processed by being dropped to the first floor to a corn meal sifting machine, and then bagged. All of this was powered by a waterwheel that would have been mounted in the wheel pit [1], and by numerous gears, driveshafts, pulleys, levers, and belts.

Thomas Shepherd's Grist Mill, Shepherdstown, West Virginia.

[TOP]
Waterwheel (c. 1891) and surge tank (c. 1905) behind the wheel.

[BOTTOM]
Waterwheel axle and hub.

Thomas Shepherd's grist mill shows how a site can change over time to adapt to new technological developments. The mill operated with traditional millstones for more than 160 years, from 1739 to the 1890s. In the latter decade, milling machinery that used rollers was added to produce a finer, better flour. Because the new machinery required more power than the old millstones, a new Fitz steel overshot waterwheel was added in the 1890s, replacing a wood wheel. The new wheel originally was located 150 feet downstream from the mill, where it could obtain the maximum water power from the nearby Town Creek, and it transmitted power by an endless-wire cable. In 1905, the iron wheel was moved next to the mill, and a surge tank was added to the end of the penstock, above the wheel, to decrease pressure on the penstock when the gate was closed.

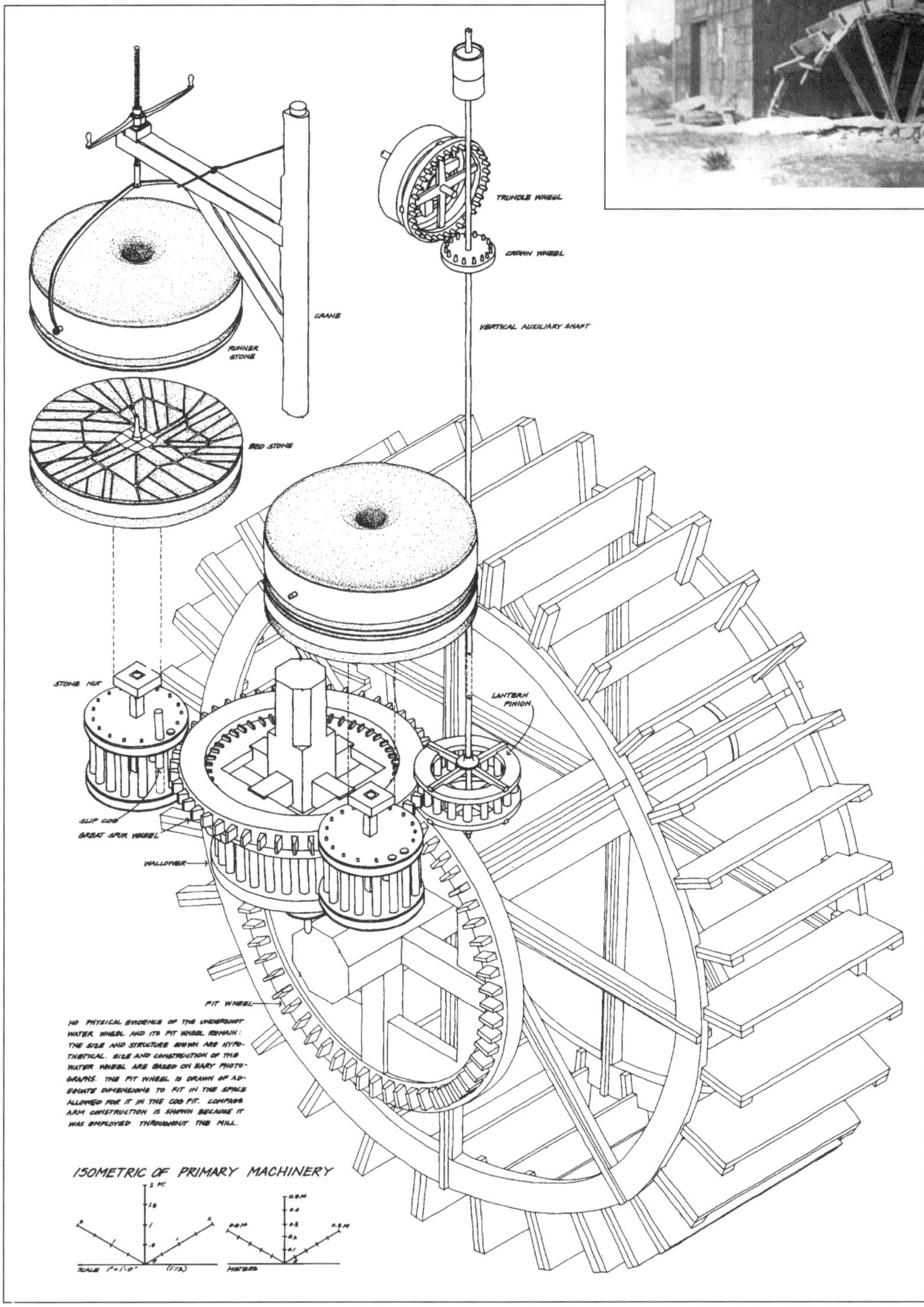

Lefferts Tide Mill, c. 1795, Huntington, New York.

[ABOVE] *Lefferts Tide Mill, c. 1920.*

[LEFT] *Primary machinery of Lefferts Tide Mill.*

This Long Island grist mill was powered by tidal flow instead of a stream. Water at high tide was impounded and then released at low tide to power the undershot waterwheel. Through wooden gearing, the waterwheel was capable of powering two millstones plus a third shaft for auxiliary equipment.

Roller Mill, Fisher-Fallgatter Mill, 1884, Waupaca, Wisconsin.

A revolution in flour milling occurred in America in the 1880s as roller mills replaced the traditional stones in the grinding process. Roller mills produced a much finer (and whiter) flour, which was more desirable to consumers, than did stone-grinding. Following the development of the roller mill process in Hungary, John Stevens, an inventor in Neenah, Wisconsin, approximately 30 miles from Waupaca, is credited with developing the first practical American roller mill. Stevens received a patent in 1880 for his invention. As large midwestern mills adopted the new process, small mills all over the country were forced to adapt or perish. A "veritable stampede" was on, according to one historian. Whereas perhaps only two mills in the entire nation had eliminated their stones by 1880, the 1890 census showed that more than 25% of all flour in the United States was ground without stones.[17]

Section of Hogan Grain Elevator, 1861-62, Seneca, Illinois.

As America's economy expanded in the nineteenth century, new market opportunities appeared for farmers. Storage and transportation facilities were devised, for example, for the shipment of surplus grain to market. Small "country" elevators, such as the Hogan and Chokio grain elevators, were built throughout the Midwest to collect grain from local farmers and forward it to larger terminal elevators in urban centers. These elevators operated much the same as individual mills: grain was first elevated to the top of the structure and then directed into various hoppers and bins by gravity. The bucket conveyor belt that elevated the grain in the Hogan Grain Elevator originally was operated by a rope drive powered by a steam engine. From the bins, grain was directed down spouts for shipment to terminal elevators. The Hogan Grain Elevator, situated on the Illinois and Michigan Canal, had separate spouts for loading to either canal barge or railroad car.

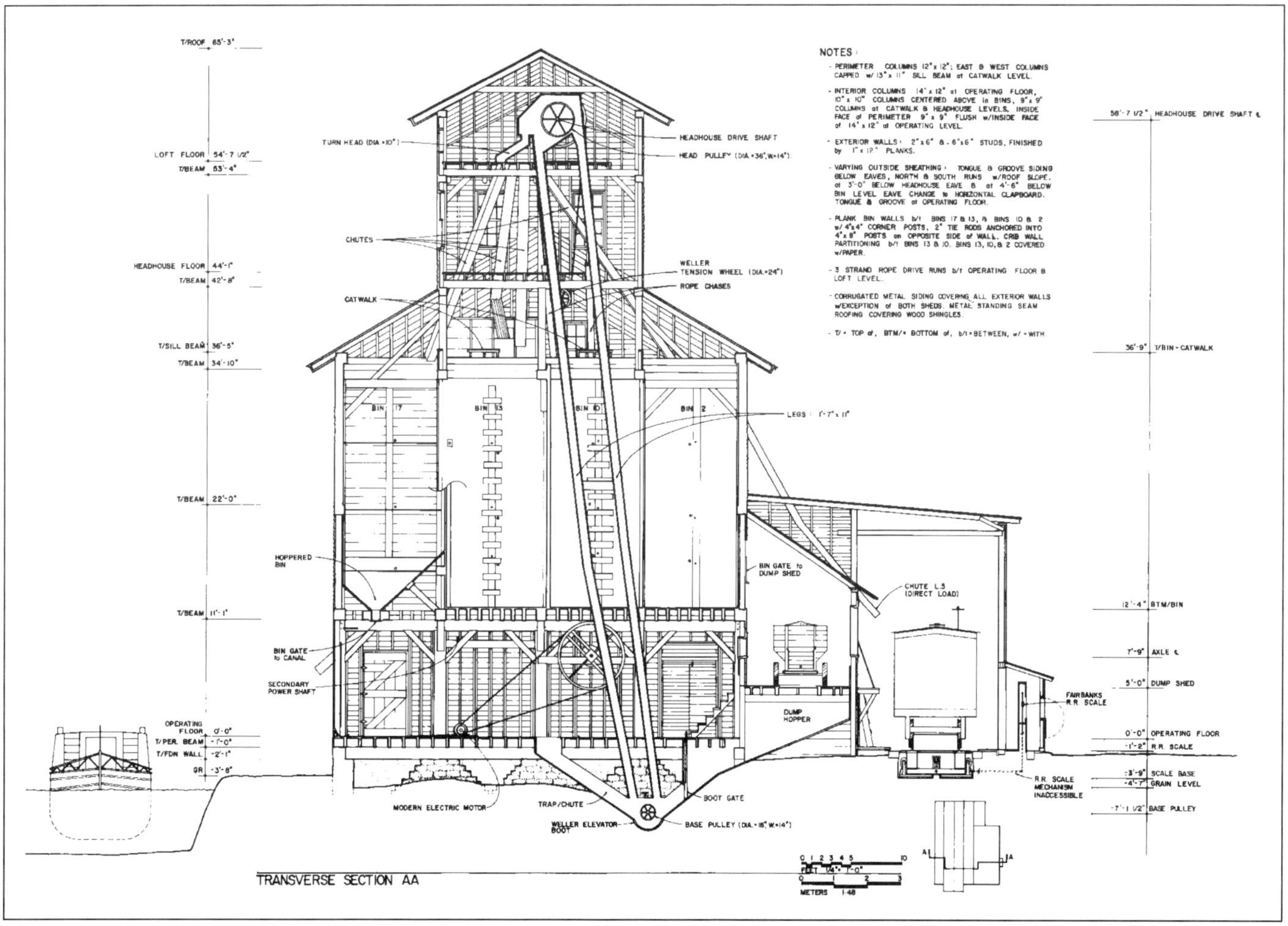

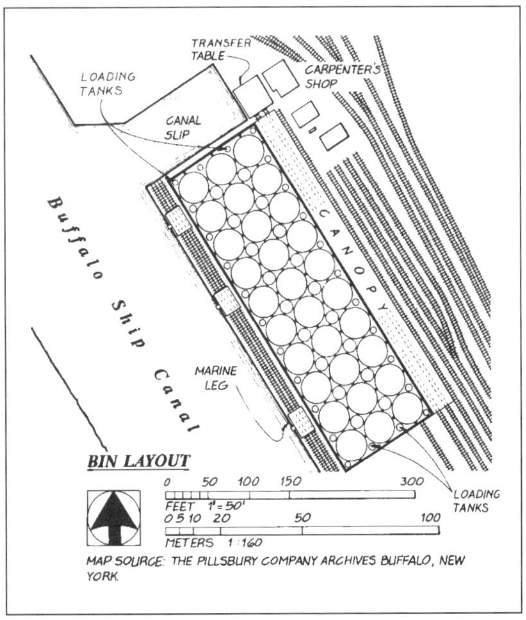

Great Northern Grain Elevator, 1897, Buffalo, New York.

[ABOVE] *The exterior of the elevator by the Buffalo Ship Canal.*

[LEFT] *Arrangement of storage bins.*

The Great Northern Grain Elevator was a terminal elevator built in Buffalo, New York, by the Great Northern Railroad in 1897. When completed, the elevator was one of the most technologically advanced in the world. The Great Northern was the first to use steel grain bins on a large scale and to apply electricity as a source of power. The use of individual electric motors allowed more interior space to be used for grain storage by eliminating cumbersome power transmission equipment. The arrangement of bins, furthermore, into main, inner-space, and outer-space bins was an efficient form later adapted for concrete elevators.

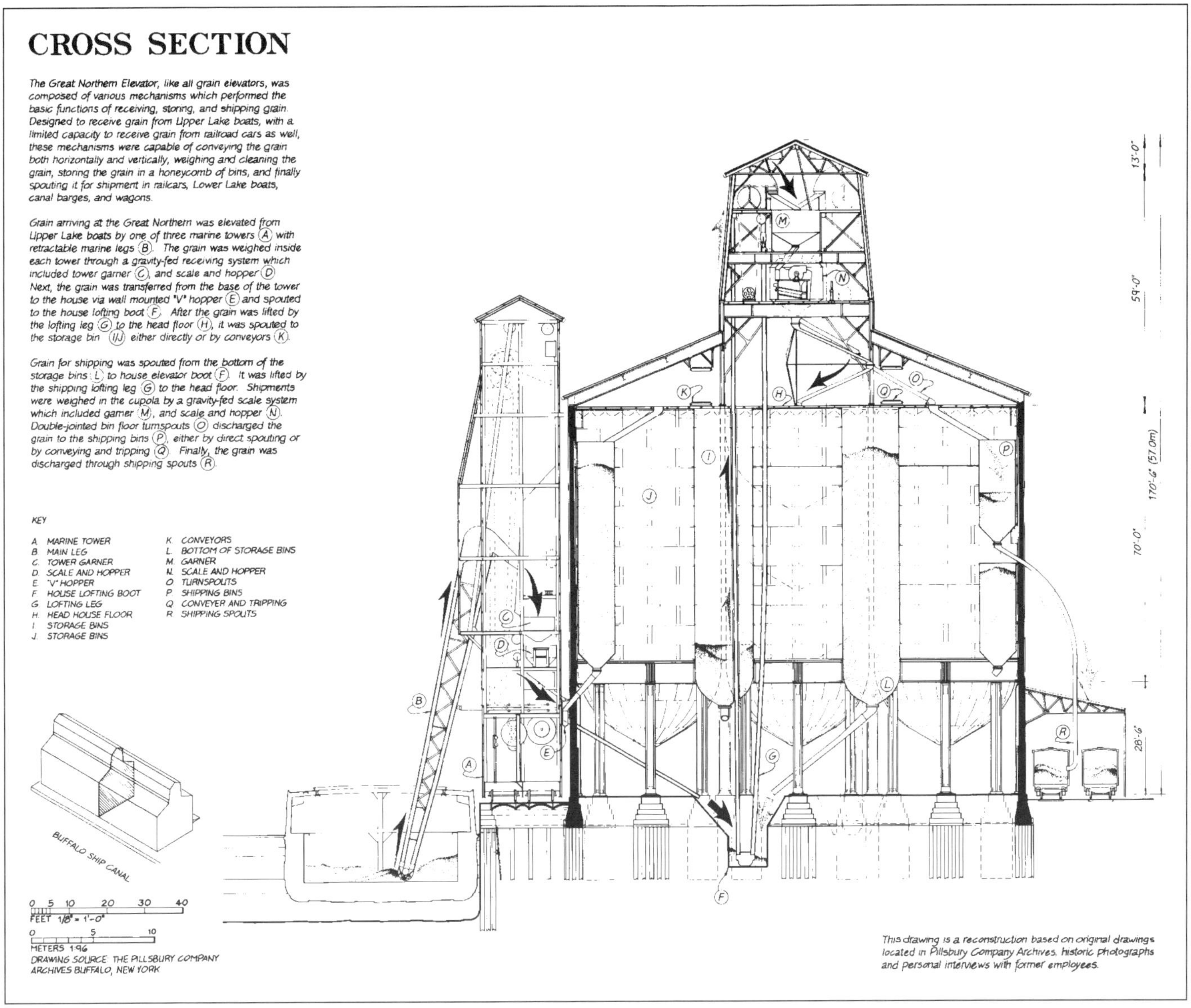

[ABOVE] *Section of Great Northern Grain Elevator.*

[RIGHT] *Bottom of the main silos in the basement level, with distribution chutes.*

The Great Northern Elevator had a capacity of 2.5 million bushels of grain and could receive grain from boats from the Great Lakes and from railroad cars. From the storage bins, grain could be spouted for shipment to railroad cars, canal barges, boats on the Great Lakes, or even wagons.

A CLOSER LOOK

Long Island Windmills

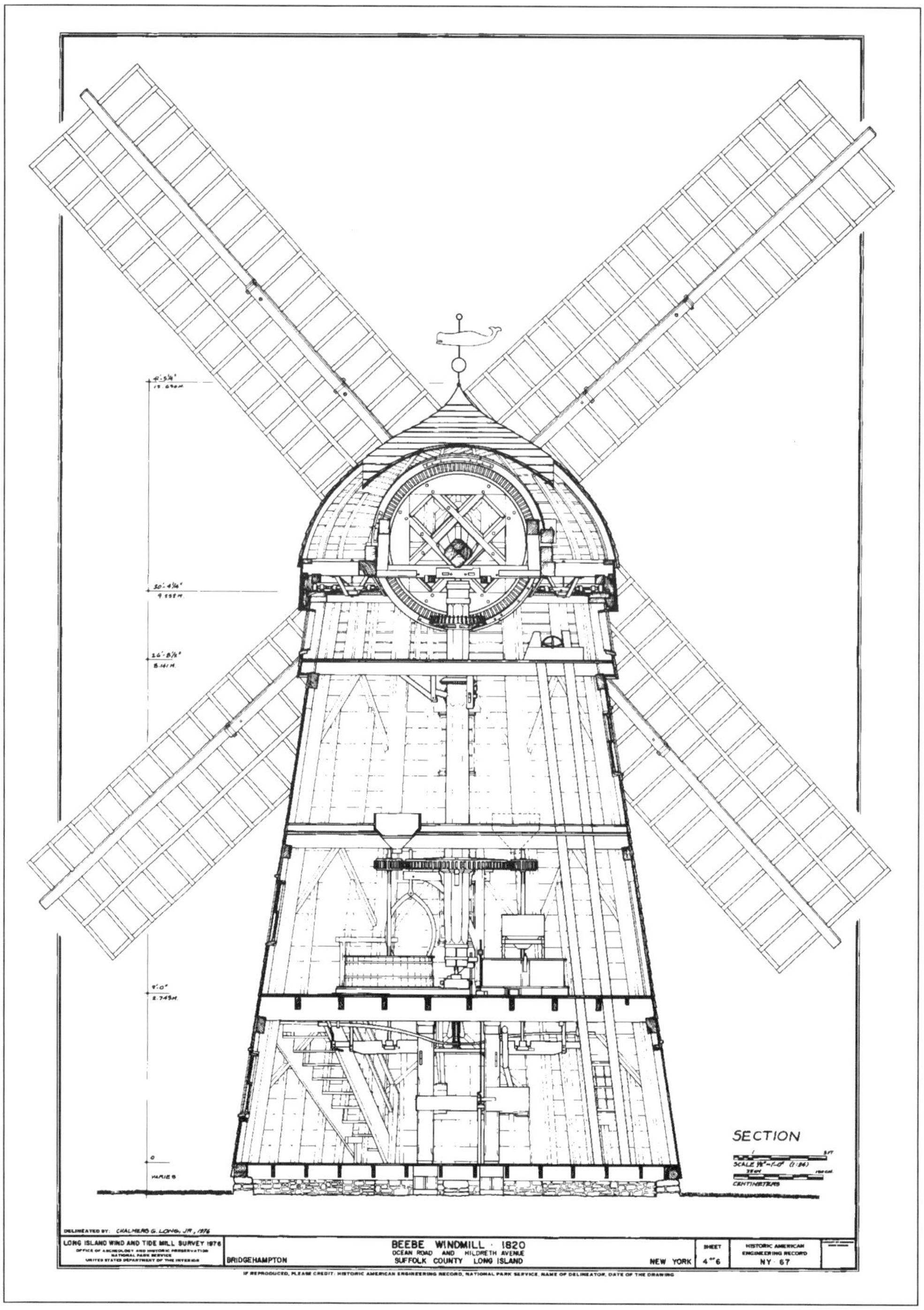

Section of Beebe Windmill, 1820, Bridgehampton, New York.

In addition to grist mills powered by water, a few were operated by wind. Several of these wind-driven grist mills, dating to the late eighteenth and early nineteenth centuries, have survived on eastern Long Island, where water power sufficient to run a mill was scarce. Most of these mills were custom mills, operating only seasonally to grind the grain of local farmers for feed. By the end of the nineteenth century, a shift in local agriculture from grain to market-garden produce, the rise of larger steam-powered mills, and the overall decline in the region's agriculture as wealthy urbanites purchased farms for vacation estates transformed most of eastern Long Island's windmills from productive enterprises into historic relics.

A CLOSER LOOK

Long Island Windmills

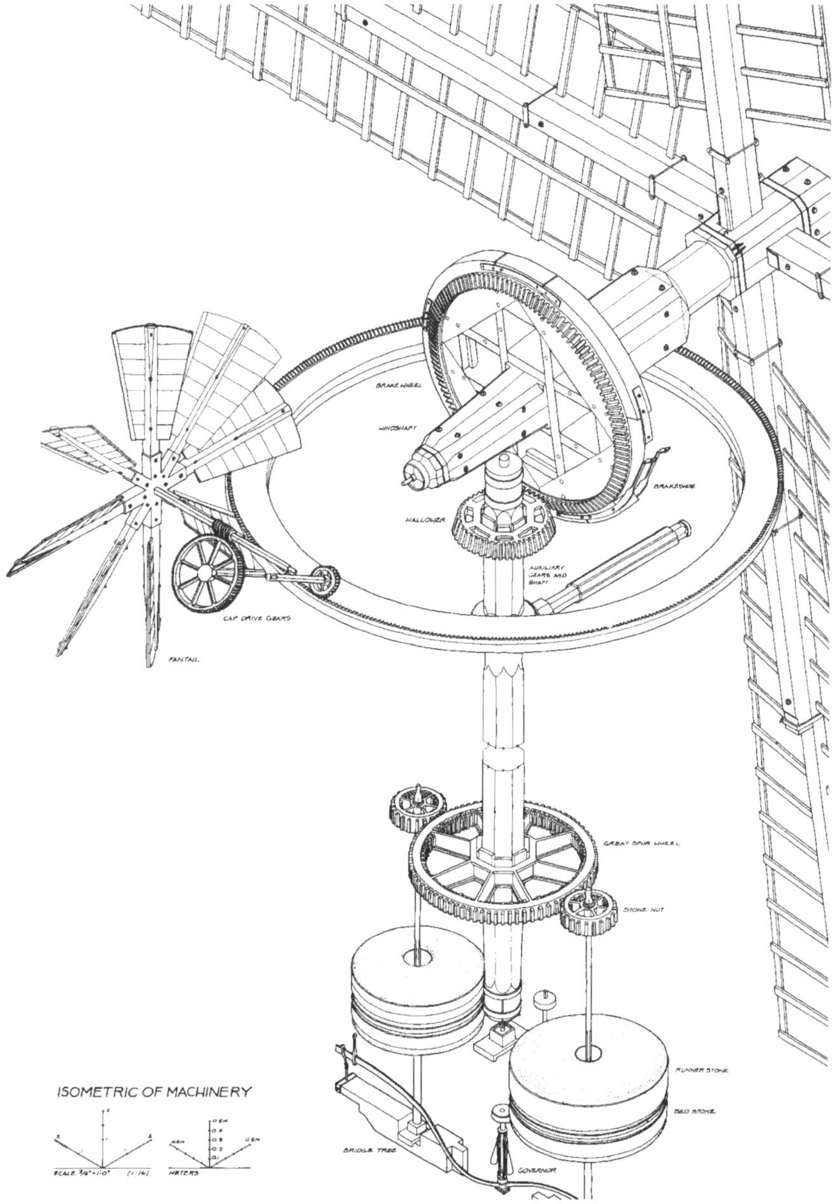

[ABOVE] *Door and* [BELOW] *stairway, Pantigo Windmill, 1804, East Hampton, New York.*

Both the Beebe and Pantigo windmills were built by Samuel Schellinger, a local craftsman who also made furniture, repaired boats, and built houses and barns as well as mills. Schellinger's later work on the Beebe Windmill (built in 1820) shows familiarity with contemporary English windmill practice and demonstrates his evolution as a millwright. A fantail to keep the sails of the mill into the wind, cast-iron bevel gearing, clasp-arm gear construction, a governor, and a decorative ogee cap all mark the Beebe Windmill as the most sophisticated Long Island mill of its day. Several of the features that made the Beebe Windmill the most sophisticated of the early Long Island windmills are evident in this machinery isometric, including the fantail, cast-iron bevel gearing, clasp-arm gear construction, and a governor located next to the millstones.

[ABOVE] *Machinery isometric of Beebe Windmill, Bridgehampton, New York.*

A CLOSER LOOK

Long Island Windmills

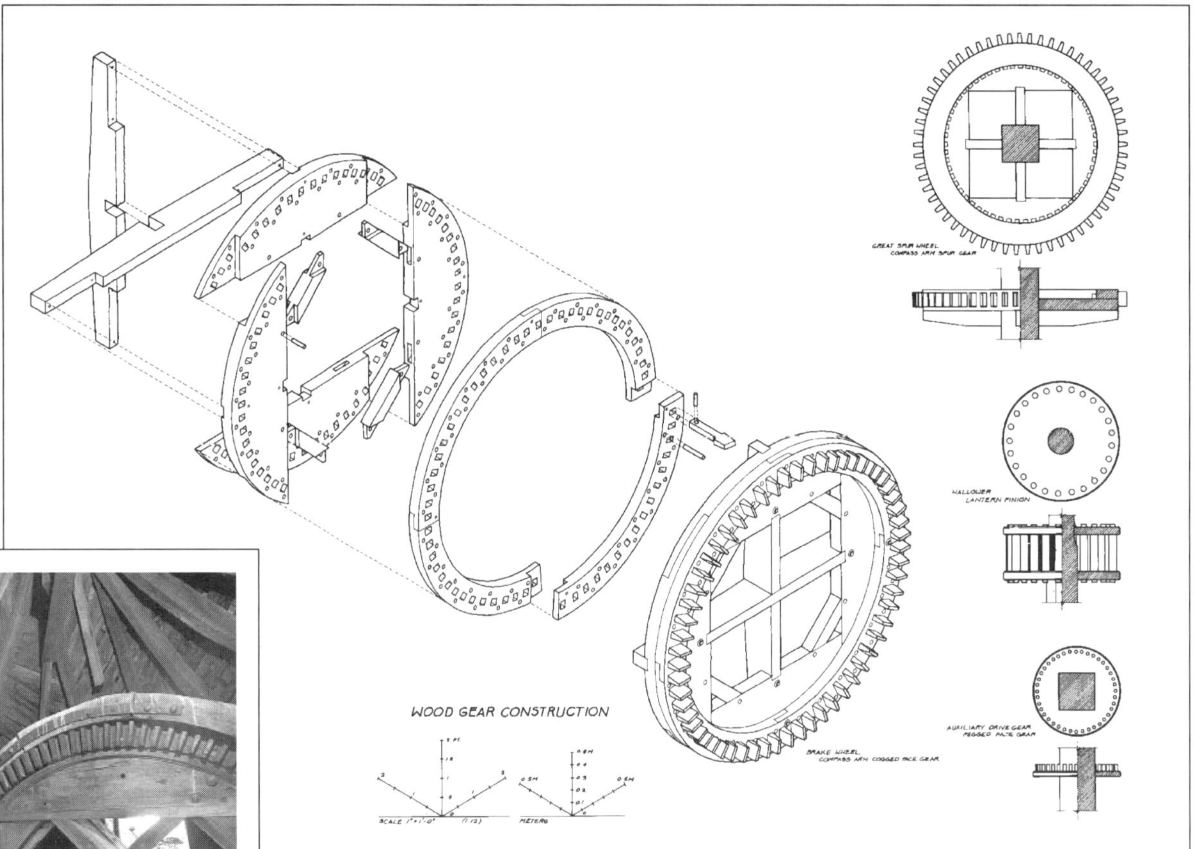

[ABOVE] *Wooden gear construction, Windmill at Water Mill, c. 1800, Water Mill, New York.*

[FAR LEFT] *Brake wheel, with cast-iron teeth, and windshaft and interior cap construction, Beebe Windmill, 1820, Bridgehampton, New York.*

[LEFT] *Wooden brake wheel, windshaft, and interior cap construction of Hook Windmill, 1806, East Hampton, New York.*

Most of the surviving Long Island windmills, with the exception of the Beebe Windmill, were constructed with wooden gears and machinery. The gears from the windmill at Water Mill, New York, demonstrate the complexity of construction of these wooden gears. The Beebe Windmill shows more advanced windmill construction technique in the use of iron gear teeth and in the clasp-arm construction of the brake-wheel gear. The arms of the wheel attach to the windshaft by enclosing the shaft; the radiating arms of the Hook Windmill brake wheel, by contrast, are mortised through the shaft. Clasp-arm construction produced a stronger gear and one that was easier to install and remove.

Drawing from Beneath: Mining and Oil

Palmetto Gulch Ten-Stamp Gold Mill, c. 1870-1900, South Pass City, Wyoming.

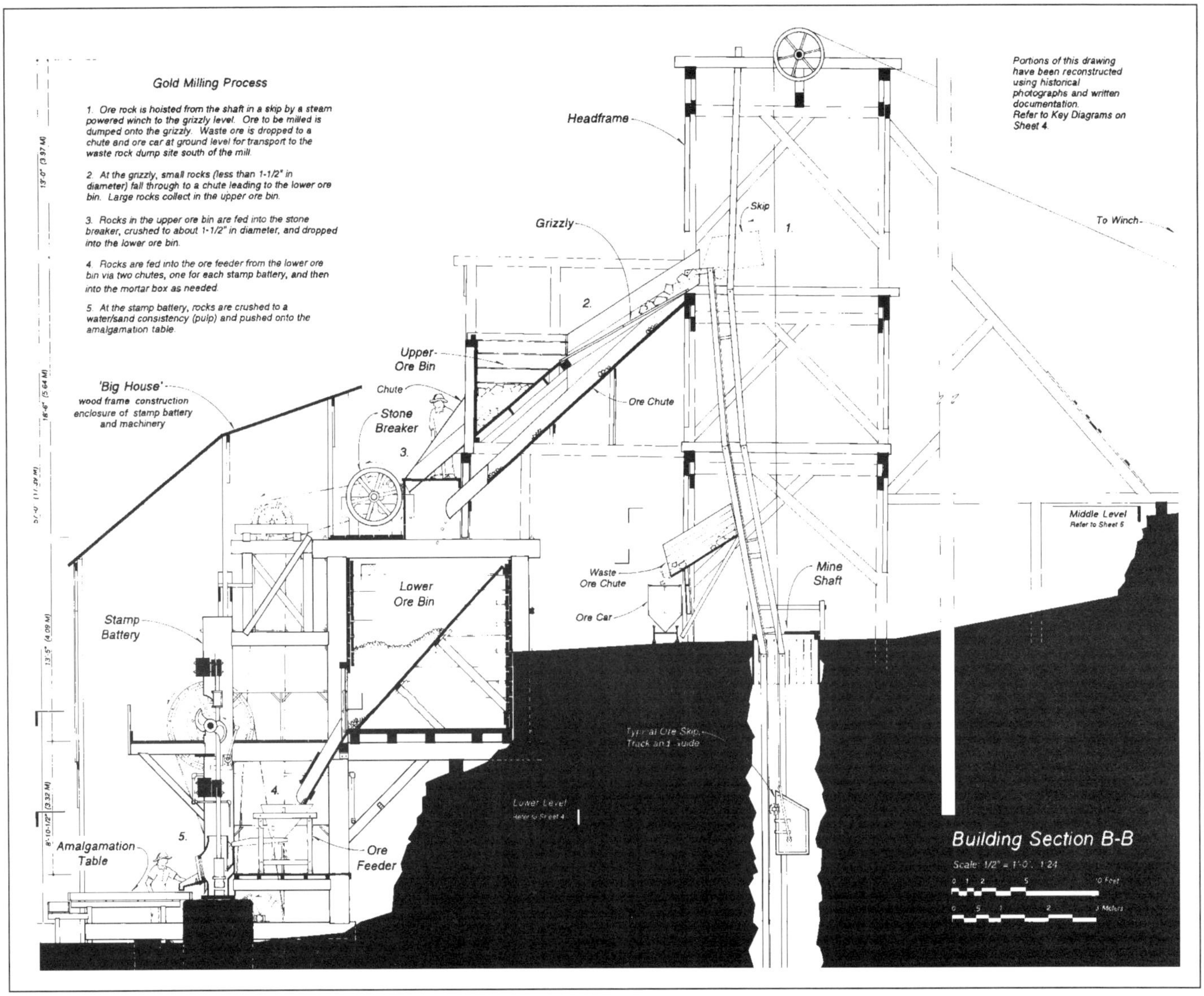

Gold Milling Process, Joshua Tree National Monument, Indio, California.

Gold had been mined in Georgia and North Carolina early in the nineteenth century, but the discovery of gold in California in 1848 produced a mining frenzy never before seen in America. At first, gold was easily obtained as loose particles in placers. As these sites diminished, miners began wresting gold ore from quartz rock veins found in the western mountains. The stamp mill, with iron or steel pounding "pestles" powered by steam or water, was invented to crush the ore into a powder, which when mixed with mercury, released the gold. The stamp mill was of enormous importance in mining areas around the world. One enthusiastic mining engineer, somewhat forgetful of the early automated grain mill, wrote in 1899 that "there is no more perfect instance of automaticity than a fully-equipped gold stamp-mill; and I cannot help thinking that it was the example of the gold stamp-mill that, more or less, influenced our manufacturers of iron and steel and other metals to introduce the same principle into their works, until to-day, machinery, rather than manual labor, effects, from first to last, all our great metallurgical operations."[18]

[LEFT] *Steel headframe, 1898-1902, Stewart Mine, Butte, Montana.*

Gold was discovered in Butte, Montana, in 1864, but silver and then copper made Butte famous in the late nineteenth century. Headframes and hoists hauled men to and from deep underground mines and brought the ore to the surface. This headframe was one of the earliest of steel in the Butte mining district, replacing a smaller wooden headframe. Ore brought to the surface in 5-ton skip cars was dumped into a chute located on the right of the headframe and shunted to loading bins for loading into railroad cars.

[BELOW] *Quincy Mining Company No. 6 Shaft-Rockhouse, 1892, Hancock, Michigan. No. 6 Shaft-Rockhouse in 1948.*

The California Gold Rush was not the only mining boom in America in the 1840s. The discovery of copper on the Upper Peninsula of Michigan competed with western gold for the attention of miners and investors. The Quincy Mining Company in Hancock, Michigan, was formed in 1846 and became one of the world's leaders in copper production, earning the nickname "Old Reliable" for its steady output. When Quincy Mining Company's mines closed in 1945, the company had produced 848 million pounds of copper. The No. 6 Shaft-Rockhouse, built in 1892 as the company expanded its operations, solved a technological bottleneck. Prior to 1892, underground copper-bearing rock was brought to the surface through a shaft house and was then transported on a tram road to a central rock house, where the rock was crushed into smaller sizes. The new structure, with five rock crushers powered by steam, was built as a combination shaft house and rock house, reducing the number of times the rock had to be handled. As in the western gold stamp mills, materials were hoisted to the top of the structure and, through gravity, were sorted among bins, crushers, and other machinery, and were finally deposited into railroad cars for further processing elsewhere.

[BELOW] *Pits and* [FACING PAGE, TOP] *mineyards of Diamond Mine, near Butte, Montana.*

Even though American industry in the nineteenth century proved remarkably resourceful in figuring out how to reuse what was previously considered "waste" products, there still accumulated an enormous amount of detritus from the industrial process, such as these 20-foot-high coal tailings from a mine in Alabama. Open-pit mines and other mining methods also radically altered the landscape, such as that around Butte, Montana.

[LEFT] *Tailings pile, Pratt Mines, Birmingham, Alabama.*

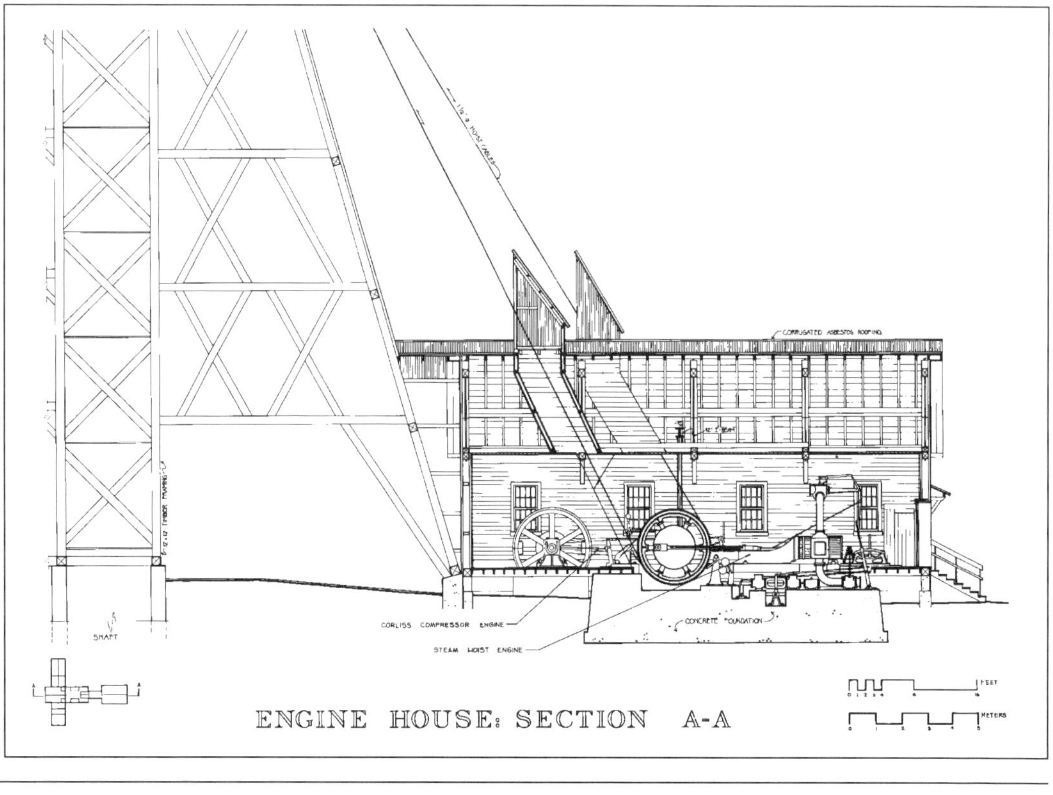

Avery Island Salt Works, Avery Island, Louisiana.

[ABOVE] *Breaker building, 1899.*

[LEFT] *Engine house, with hoist cables leading to the breaker building.*

Avery Island Salt Works, Avery Island, Louisiana.

[RIGHT] *Steam hoist engine, 1899, and cable drums in the engine house.*

[BELOW] *Stockpile of salt 900 feet underground.*

The first rock-salt mine in America was sunk on Avery Island, Louisiana, in 1867. Salt from the site had been used since 1000 A.D., through surface excavation and evaporation techniques. Large-scale production began during the Civil War, when the Confederate government took possession of the site and opened additional pits. Poor engineering made the original mine unworkable by the late 1890s. At the same time, however, America became the world's leader in salt production, and a new shaft and surface structures were constructed by the Avery Rock Salt Mining Company in 1898 and 1899. The breaker building was built in 1899 by E.H. Park of New Orleans, Louisiana. A timber-framed structure, the building was designed with machinery to bring rock salt from below ground and, through a gravity process, crush chunks of salt into small pieces, grind these pieces into finer particles, and store, package, and load the salt for shipment. Rock salt was brought to the surface by a steam hoisting engine, built in 1899 by Vulcan Iron Works of Wilkes-Barre, Pennsylvania, and perhaps the oldest operating engine of its kind.

West Oil Company, c. 1895, Volcano, West Virginia.

[RIGHT] *Oil derrick.*

[BELOW] *Endless-wire cables running between an engine house and an oil pump.*

Before electricity was widely available in such places as West Virginia's oil fields, ingenious methods were devised to power the pumps that brought oil to the surface. Instead of placing a power source, such as a gas engine, next to each well, the West Oil Company near Volcano, West Virginia, ran endless-wire cables from one 20-horsepower gas engine to all of its wells. Each cable turned the large crank wheel of an oil rig, which rocked a horizontal walking beam up and down, which in turn powered a pump's piston rod. Linear motion was thus converted to rotary motion that finally was transformed into vertical pumping motion.

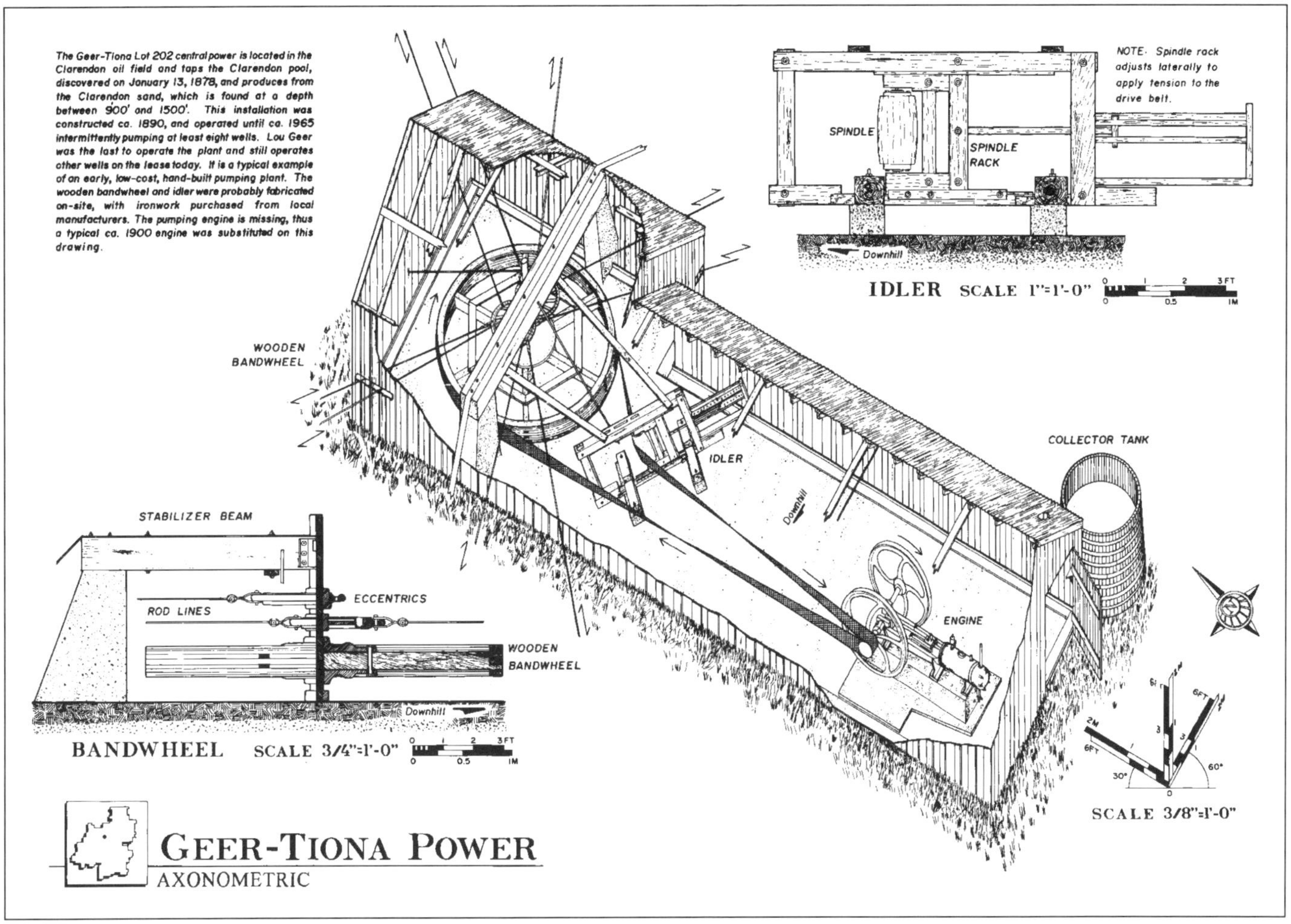

Geer-Tiona Central Power, c. 1890, near Warren, Pennsylvania.

This wooden bandwheel and a gas engine similar to the one depicted powered cables attached to individual oil rigs in the Clarendon oil field of Allegheny National Forest in Pennsylvania.

Urban Necessities: Water and Gas

North Point Water Tower, 1874, Milwaukee, Wisconsin.

[ABOVE] *Interior, showing spiral staircase and wrought-iron standpipe.*

The column of water in the standpipe maintained a constant pressure in the city's water mains, regardless of changes in demand.

[FACING PAGE, LEFT] *Ornate architecture of the water tower.*

[FACING PAGE, RIGHT] *Section drawing showing wrought-iron standpipe.*

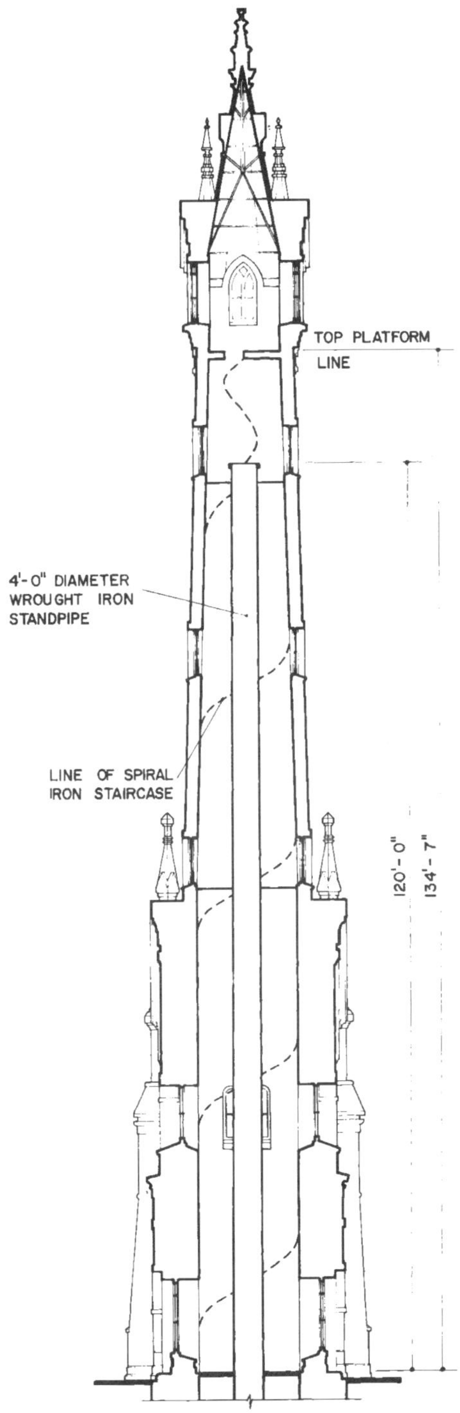
TOP PLATFORM
LINE
4'-0" DIAMETER
WROUGHT IRON
STANDPIPE
LINE OF SPIRAL
IRON STAIRCASE
120'-0"
134'-7"

SECTION A-A

[FACING PAGE]

Pumping Station No. 1, 1856, Louisville, Kentucky.

The elaborate Pumping Station No.1 and its associated standpipe were designed and built in 1856 by Theodore R. Scowden and Charles Hermany of the Louisville Water Company. In the 1870s, Hermany carried out important experiments here in rapid sand filtration of water, a method that became standard in later urban water purification systems. The standpipe is 169 feet high and was rebuilt by the city after being toppled by a tornado in 1890.

[THIS PAGE]

[TOP RIGHT]
North Grand Water Tower, 1871, St. Louis, Missouri.

[TOP LEFT]
Roxbury Standpipe, 1870, Boston Water Works, Boston, Massachusetts.

[BOTTOM RIGHT]
Compton Hill Water Tower, 1898, St. Louis, Missouri.

[BOTTOM LEFT]
Chicago Water Tower, 1869, Chicago, Illinois.

Elysian Water Tower, 1895, Elysian, Minnesota.

New Croton Dam and Reservoir, 1892-1907, near Croton-on-Hudson, New York.

The New Croton Dam was designed to provide more water to New York City as that metropolis rapidly expanded in the late nineteenth century. The dam is a masonry gravity dam, with a maximum height of almost 300 feet, which made it the tallest dam in the world when completed.

Troy Gasholder House, 1873, Troy, New York.

The Troy Gasholder House is one of the most elaborate of surviving gasholder houses. Built in Troy, New York, in 1873, the gasholder house enclosed an expandable iron tank that stored gas principally for lighting. When empty, the telescoping tank rested in a pit of water, which acted as a seal. The brick structure protected the gasholder in inclement weather and helped prevent water in the pit from freezing. The roof of the structure and the ventilating cupola are supported by wrought-iron trusses. [19]

[ABOVE] *Exterior*

[BELOW LEFT] *View of roof trussing from below.*

[BELOW RIGHT] *Lower-chord tie rods of roof trussing.*

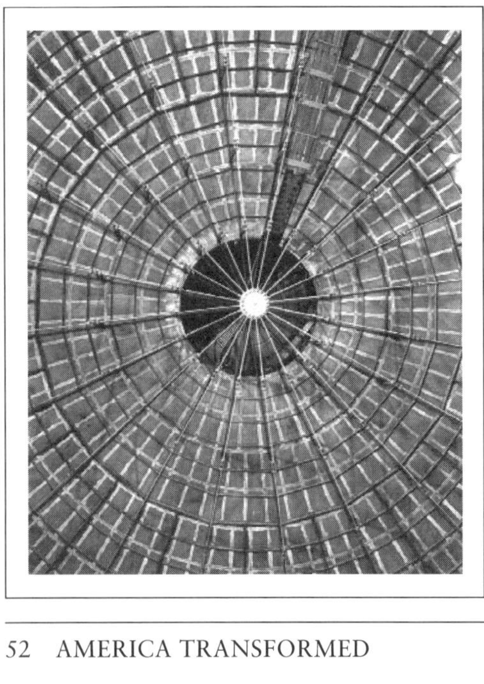

Concord Gasholder House, 1888, Concord, New Hampshire.

[ABOVE] *Gasholder house.*

[ABOVE, RIGHT] *Gasholder guide wheel.*

[RIGHT] *Interior, showing gasholder guide wheels and rails and top of empty gasholder tank.*

The Concord Gasholder House, built in 1888, still retained its original gasholder when recorded by HAER in 1982. The gasholder was constructed of riveted wrought-iron plates and had a capacity of 120,000 cubic feet of gas. Iron wheels and guide rails guided the gasholder as it expanded and contracted according to the volume of gas within.

Petersburg Gasholder, 1876, Petersburg, Virginia.

[LEFT] *Guide wheels and base supporting columns.*

[BELOW] *Cast-iron frame of gasholder.*

As urban and therefore public engineering, even gasholders may acquire a certain stylish quality. The Petersburg Gasholder, built in 1876 by the Richmond, Virginia, firm of Tanner and Company, featured tasteful columns and decorative, perforated top beams, all of cast iron. Gasholders in southern climates did not need all-enclosing structures.

References

1. Robert H. Wiebe, *The Search for Order, 1877–1920* (New York: Hill and Wang, 1967) p. 15.

2. Carroll Pursell, *The Machine in America: A Social History of Technology* (Baltimore: Johns Hopkins University Press, 1995) p. 119.

3. Louis C. Hunter, *A History of Industrial Power in the United States, 1780–1930, Vol. I: Waterpower in the Century of the Steam Engine* (Charlottesville, VA: University Press of Virginia, 1979) p. 38.

4. Ruth Schwartz Cowan, *A Social History of American Technology* (New York: Oxford University Press, 1997) p. 72.

5. John W. McGrain, "'Good Bye Old Burr': The Roller Mill Revolution in Maryland, 1882." *Maryland Historical Magazine*, 77 (Summer, 1982):165–167.

6. Alan I. Marcus and Howard P. Segal, *Technology in America: A Brief History* (San Diego: Harcourt Brace Jovanovich, 1989) pp. 196–199.

7. David Wasson, "The Modern Type of Oppression." *North American Review* (1874), as quoted in Alan Trachtenberg, *The Incorporation of America: Culture and Society in the Gilded Age* (New York: Hill and Wang, 1982) p. 113.

8. Bernard Bailyn, et al., *The Great Republic,* Vol. 2, 3rd Ed. (Lexington, MA: D.C. Health, 1985) p. 546.

9. David E. Nye, *American Technological Sublime* (Cambridge, MA: MIT Press, 1994) p. 118.

10. Trachtenberg, *The Incorporation of America*, p. 114.

11. Thomas C. Cochran and William Miller, *The Age of Enterprise: A Social History of Industrial America* (New York: HarperCollins, 1961, revised edition) p. 251.

12. Louis C. Hunter, *A History of Industrial Power in the United States, Vol. II: Steam Power* (Charlottesville, VA: University Press of Virginia, 1985) p. 515.

13. John B. Rae, "Energy Conversion," in Kranzberg, Melvin, and Carroll W. Pursell, Jr., eds., *Technology in Western Civilization, Vol. I* (New York: Oxford University Press, 1967), pp. 347–348.

14. Brooke Hindle and Steven Lubar, *Engines of Change: The American Industrial Revolution, 1790–1860* (Washington, D.C.: Smithsonian Institution Press, 1986) p. 46.

15. John Michael Vlach, *Back of the Big House: The Architecture of Plantation Slavery* (Chapel Hill, NC: University of North Carolina Press, 1993) pp. 10–11, 125–127.

16. Richard O'Connor, "Magnolia Plantation Cotton Gins and Presses," U.S. Department of the Interior, Historic American Engineering Record (HAER), No. LA-11, Prints and Photographs Division, Library of Congress, Washington, D.C., 1997, p. 9.

17. Polly Athan, "Technological Watersheds on the Waupaca River: A History of the Fisher-Fallgatter Mill," U.S. Department of the Interior, Historic American Engineering Record (HAER), No. WI-1, Prints and Photographs Division, Library of Congress, Washington, D.C., 1979, p. 14; McGrain, "Good Bye Old Burr," p. 157.

18. *Transactions of the American Institute of Mining Engineers*, 29 (1899), lii, quoted in Pursell, *The Machine in America*, p. 163.

19. Theodore Anton Sande, *Industrial Archeology: A New Look at the American Heritage* (New York: Penguin, 1976) pp. 58–63.

2

Making Things

Thomas Jefferson was on the wrong side of Oliver Evans, and on the wrong side of history, in 1813. Thirty years earlier, in the 1780s, Evans had devised a system of milling in which grain was automatically transported throughout the mill from one operation to the next. Evans's ideas constituted a revolution in materials handling that predated by decades similar developments in other industries. In the early years of the nineteenth century, however, it was not at all clear the direction American industrialization would take, or even if the young nation would industrialize at all. Even visionaries such as Jefferson could not foresee the changes ahead. Disgruntled millers in Baltimore, Maryland, appealed to Jefferson in 1813, after they had petitioned the U.S. Congress not to renew Evans's patent rights to his automatic milling machinery, claiming Evans had not really "invented" anything. Jefferson, although a famous advocate for American agrarianism, was also fascinated by technology and was an inventor himself. But he failed to see Evans's pioneering work as a whole, and he replied in a letter to the millers that Evans's "elevator was nothing more than the old Persian wheel of Egypt and that the conveyor is the same as the screw of Archimedes..."[1] Although Congress did extend his patent rights in 1814, Evans bitterly resented Jefferson's letter. There was no way for Jefferson to have realized it, but Evans had achieved, according to Siegfried Giedion, "what was to become the pivot of later mechanization." Evans was simply ahead of his time, and it took most of the nineteenth century for the rest of the country to catch up.[2]

Increasing mechanization and automation were only two elements, albeit very important ones, contributing to the varied patterns of American industrialization in the nineteenth century. The rise of the factory, the transfer of European technical knowledge, the development of a skilled workforce, the quest for interchangeable parts manufacturing, the patent system, the formation of large corporations, the creation of company paternalism, the evolution of scientific management, the adoption of newer power sources, and still other factors were all part of industrialization. However, not all industries and technologies were equally affected by all of these elements, and the timing of industrialization for various sectors of the economy varied drastically. Handwork and small-scale production also continued to play a significant role. Firearms, textile, iron and steel, and meat-processing industries, in particular, helped set the pace for American industrialization. Textile production helped establish the early model for industrialization, and the iron and steel industry, although not devoid of small specialty firms, had emerged by century's end as a large-scale enterprise with complex vertical and horizontal linkages. But not all industries followed the patterns set by the textile factories and iron and steel mills. For a variety of reasons,

[FACING PAGE]
Carding machine, c. 1865, Watkins Woolen Mill, Lawson, Missouri.

Mill workers in one of the Amoskeag millyards, c. 1880, Manchester, New Hampshire.

certain industries such as glass and brick were slow to mechanize, and even in 1900, there were still many local craft shops that resembled the Gruber Wagon Works, in Berks County, Pennsylvania, which combined hand labor with limited mechanized production to produce a complete product, in this case wagons, from top to bottom.

American industrialization owed much of its growth in the nineteenth century to the advent of new power sources, particularly steam technology, because it freed factories from the geographic constraints of water power. Water was the primary power source for the first half of the nineteenth century and still accounted for about one-half of all industrial power in the country in 1860. The extensive use of water in antebellum America is illustrated by the Blackstone River. This small river flows for only a little more than 40 miles from Worcester, Massachusetts, to Providence, Rhode Island. But in 1840, it powered 150 industrial plants: 94 cotton mills, 22 woolen mills, and 34 machine shops and ironworks.[3] Steam, however, quickly became the dominant energy source after the Civil War, because of its cheapness of operation and flexibility.[4] Even as early as 1832, one writer claimed that, "By the side of this tremendous energy, every other power becomes insignificant."[5] By 1899, steam accounted for more than 8 million horsepower in American manufacturing, whereas water produced only 1.5 million.[6] The actual transmission of power in factories was achieved at first by direct drive gearing, but after the 1820s primarily by leather belt drives.[7] "We can scarcely step into a shop or factory of any description," declared a contemporary observer, "without encountering a mass of belts which seem at first sight to monopolize every nook in the building and leave little or no room for anything else."[8] In the 1890s, however, the method of power transmission began to change, as factories and other users of power began to shift from generating their own power with a prime mover to being part of a larger network receiving power from a central prime mover, the power transmitted increasingly by electricity.[9]

Cotton, Cards, and Clothes: Textiles

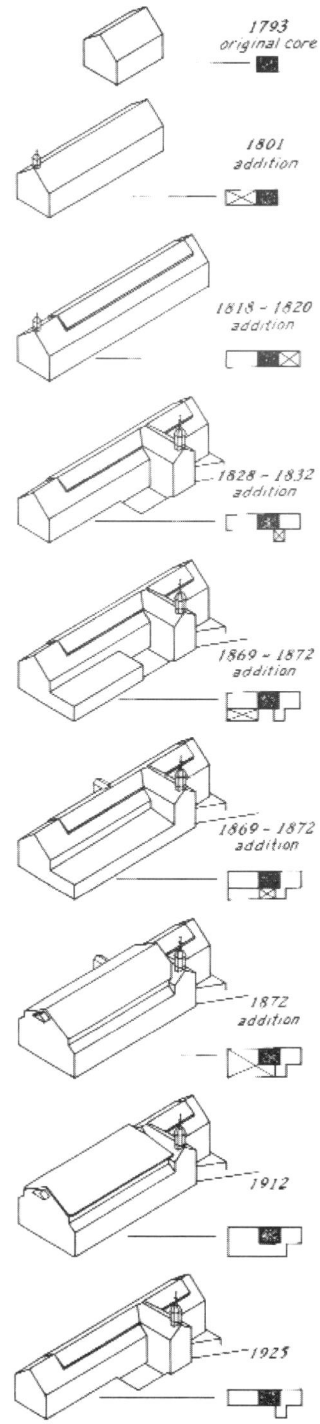

[ABOVE] *Slater Mill, 1793, Pawtucket, Rhode Island.*

[RIGHT] *Evolution of Slater Mill, 1793-1925.*

Slater Mill, built by Samuel Slater (1768-1835) in 1793, was the first successful water-powered cotton-spinning mill in America and is heralded as the birthplace of the textile industry in the United States. But Slater Mill also demonstrates how a factory can change over time, both structurally as well as in terms of what takes place in the factory. The Slater Mill building underwent several periods of alteration until 1925, when a preservation group returned the mill to what was believed to be its appearance in 1835. As Slater Mill declined in importance in relation to newer and larger mills built elsewhere, other industries moved into the building. These included machine shops, rug weavers, wood turners, coffin trimmers, jewelers, electrical suppliers, and twine manufacturers. The third floor was even used as a bicycle-riding rink in the late nineteenth century.

[ABOVE] *Lippitt Mill, 1810, West Warwick, Rhode Island.*

[RIGHT] *Detail of stairs, Monadnock Mill No.1, 1836-1839, Claremont, New Hampshire.*

These mills were typical of the small, wood-frame New England textile mills that sprang up throughout the region in the early years of the nineteenth century. Moses Brown, an early mill owner, wrote to his son in 1810, "Our people have 'cotton mill fever' as it is called. Every place almost occupied with cotton mills... spinning yarn and making cloth is become our greatest business." [10]

[FACING PAGE] *Amoskeag Mills, 1838, Manchester, New Hampshire.*

The Amoskeag Mills were enormous, stretching for more than a mile along both banks of the Merrimack River. The mills' output by century's end was claimed to be 4 million yards of cloth per week.[11]

[RIGHT] *Graniteville, South Carolina.*

[BELOW] *Graniteville Mill, 1847.*

Started by William Gregg in 1847, the Graniteville Mill was one of the South's most important early textile operations. Gregg was an influential and active proponent of southern industrialization, and he built his mill and village as models to promote his views. Gregg housed his workers in remarkable Gothic Revival-style cottages, several of which are still extant.

Cannelton Cotton Mills, 1849-51, Cannelton, Indiana.

Cannelton Cotton Mills, completed in 1851, became one of the largest producers of cotton sheeting outside of New England. The impressive sandstone-faced structure was designed by Thomas A. Tefft, a Providence, Rhode Island, architect who earlier had designed the monumental Union Station in Providence. The mill was designed to run on steam power, fueled by nearby coal deposits.

Watkins Woolen Mill, 1861, Lawson, Missouri.

[LEFT] *A row of carding machines.*

[BELOW] *Spinning jacks (or mules), each of which spun woolen yarn onto 216 spindles.*

Watkins Woolen Mill was built in Lawson, Missouri, in 1861, and produced woolen yarn and cloth until about 1900, when the doors were shut and the mill left intact. The varied textile machinery left in the mill, including, for example, two types of spinning jacks (or mules) and 17 looms of seven different types, demonstrates the variety of ways nineteenth-century machine builders addressed a technological problem. As historian Larry Gross discovered, Watkins Woolen Mill also gives "mute testimony" to the ways in which workers adapted machinery, altering less-efficient looms, for example, with an "improvised" copy of a design from a superior loom.

Harmony Manufacturing Company Mill No. 3 (Mastodon Mill), 1866-72, Cohoes, New York.

[ABOVE] *Life-size statue of Thomas Garner, principal developer of Harmony Manufacturing Company.*

[RIGHT] *One of two large Boyden water turbines (c. 1871), which powered the mill's machinery.*

The Harmony textile complex in Cohoes, New York, incorporated in 1836, was one of the largest textile enterprises outside of New England. Mill No. 3, built in 1866, was one of the most impressive architecturally. The mill became locally known as the "Mastodon Mill" after a mastodon skeleton was found during excavation for the foundation. Because of their large size and age, the two remaining 102-inch Boyden water turbines, dating to about 1871, have been designated National Historic Mechanical Engineering Landmarks.

Sibley Manufacturing Company, 1880-1882, Augusta, Georgia.

[ABOVE] *"Josiah," the company's bell, in one of the central towers (cast by the McShane Bell Foundry of Baltimore, Maryland, in 1881).*

[RIGHT] *The ornate central towers of the mill.*

As the textile industry moved to the South in the late nineteenth century in search of cheaper labor, Augusta, Georgia, became one of the region's textile centers. The mill of the Sibley Manufacturing Company was built between 1880 and 1882 on the former site of the old Confederate Powder Works. Jones S. Davis, a mill builder originally from Massachusetts, designed the building. Sibley became one of the South's leading textile firms, with more than 40,000 spindles and 1,109 looms by 1895.

Hell With the Lid Off: Iron and Steel

Principal (work) arch, Buena Vista Furnace, 1847, near Armagh, Pennsylvania.

This iron furnace, located in western Pennsylvania, operated on the same principle as colonial iron furnaces. Three raw materials went into the furnace to produce iron: iron ore, charcoal as a fuel, and some form of limestone, which helped separate impurities from the iron. Workers "charged" the furnace from the top (by way of a charging bridge, no longer extant), ignited the charcoal, and later released molten iron at the bottom, into sand molds. The combustion process depended on a blast (hence "blast" furnace) of air blown into the furnace by water-powered bellows. The Buena Vista Furnace presumably was a "cold blast" furnace, in that the blast air was not preheated.

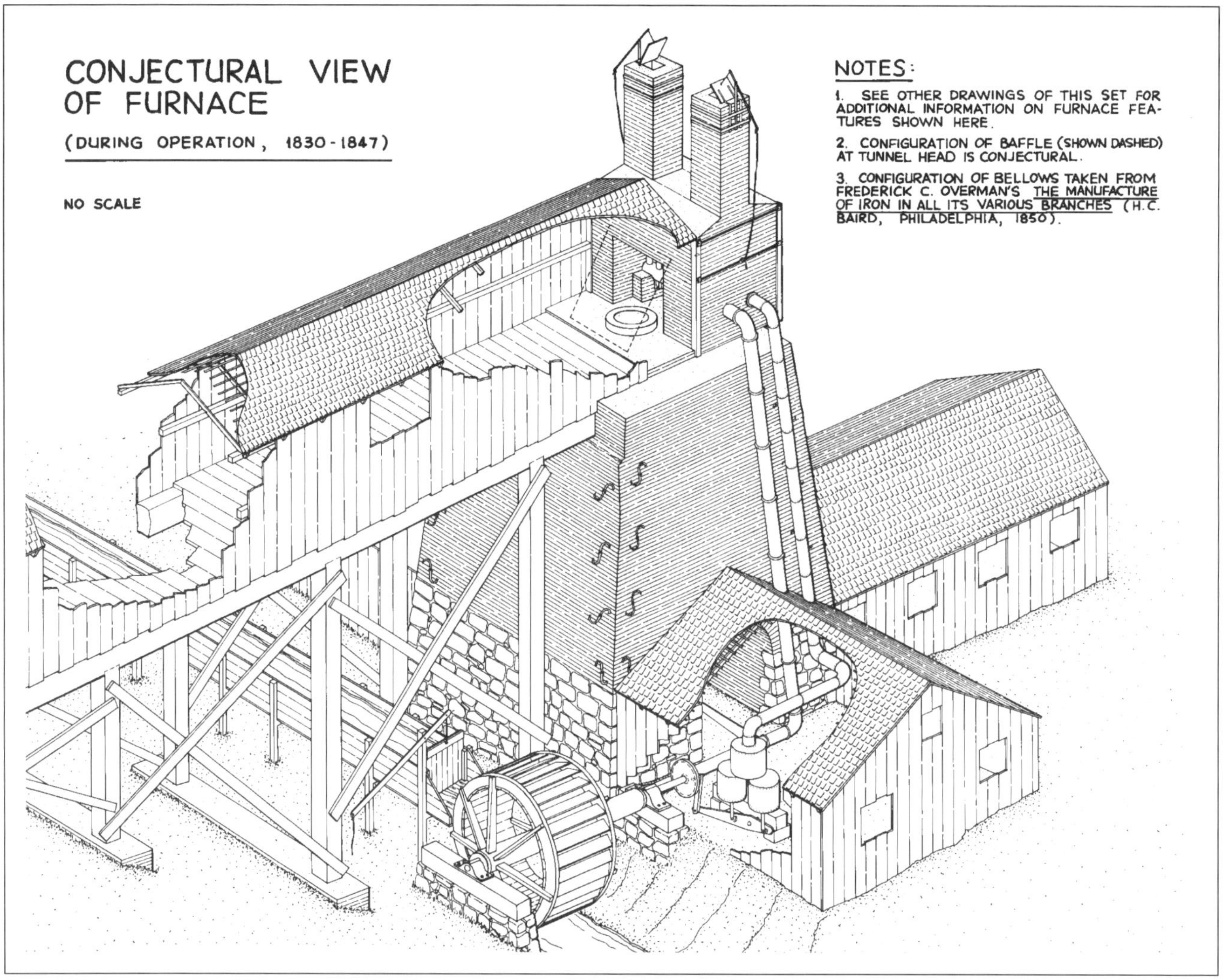

[ABOVE] *Conjectural drawing, Nassawango Iron Furnace, c. 1836-1847, near Snow Hill, Maryland.*

Despite the lonely appearance of most surviving early nineteenth-century iron furnaces, they usually were at the heart of a busy complex of machinery, structures, and corollary industries. Structures surrounding the furnace stack itself may have included, as in this conjectural view of Nassawango Furnace, a charging bridge, a headrace that brought water to a waterwheel, and bellows. In this case, the bellows blew air into a stove above the furnace that preheated the air before it entered the furnace. When molten iron was discharged from the furnace, it flowed into a cast house, shown to the right of the furnace. Furnace sites usually included several industrial structures other than the furnace itself, such as a blacksmith shop, a forge, a sawmill, a grist mill, a charcoal storage house, workers' housing, the ironmaster's house, and other community structures such as a church and a company store. Nearby were found ore mines, the vast woodlands needed to obtain wood for charcoal, and a canal, rail line, or other mode of transportation.

[FACING PAGE] *Eliza Furnace, 1845-46, near Vintondale, Pennsylvania.*

A major improvement in the efficiency of iron furnaces was the British invention in 1828 of the "hot blast" method, in which the blast air was preheated before entering the furnace, thereby increasing the efficiency of the combustion process. This not only saved charcoal fuel (by an estimated 20%), but also led to the eventual use of anthracite coal and, later, coke as fuel, both of which required higher temperatures to ignite.[12] The Eliza Furnace and the Nassawango Furnace both retain original stoves on top of the stacks that were part of the hot blast apparatus. Air was forced into the stove through a pipe from the bellows, heated, and directed back down the outside of the furnace through another pipe and into the furnace through a nozzle called a tuyere. The first hot blast stoves were separate entities placed on the ground and heated with charcoal (or coal), but ironmasters soon learned to make use of the hot waste gas emitted from the furnace opening at the top of the stack.

Adirondack Iron and Steel Company, Tahawus, New York.

The "New Furnace" of the Adirondack Iron and Steel Company, built between 1849 and 1854, made effective use of new coal-smelting blast furnace technology without actually converting to coal as fuel. This technology included more powerful blast engines (powered by waterwheels); a slightly larger, reconfigured, and heavily reinforced furnace stack; and a hot blast stove. The blowing machinery was built by the Hudson River Iron and Machine Company of Ft. Edward, New York. The charging bridge, as shown, also included an ore stamp that crushed iron ore into smaller sizes to be charged into the furnace.

[ABOVE] *The "New Furnace," 1849-1854.*

[FACING PAGE]

[TOP] *Site section, the "New Furnace."*

[BOTTOM] *Waterwheels (conjectural) and blowing cylinders that provided the air blast for the "New Furnace."*

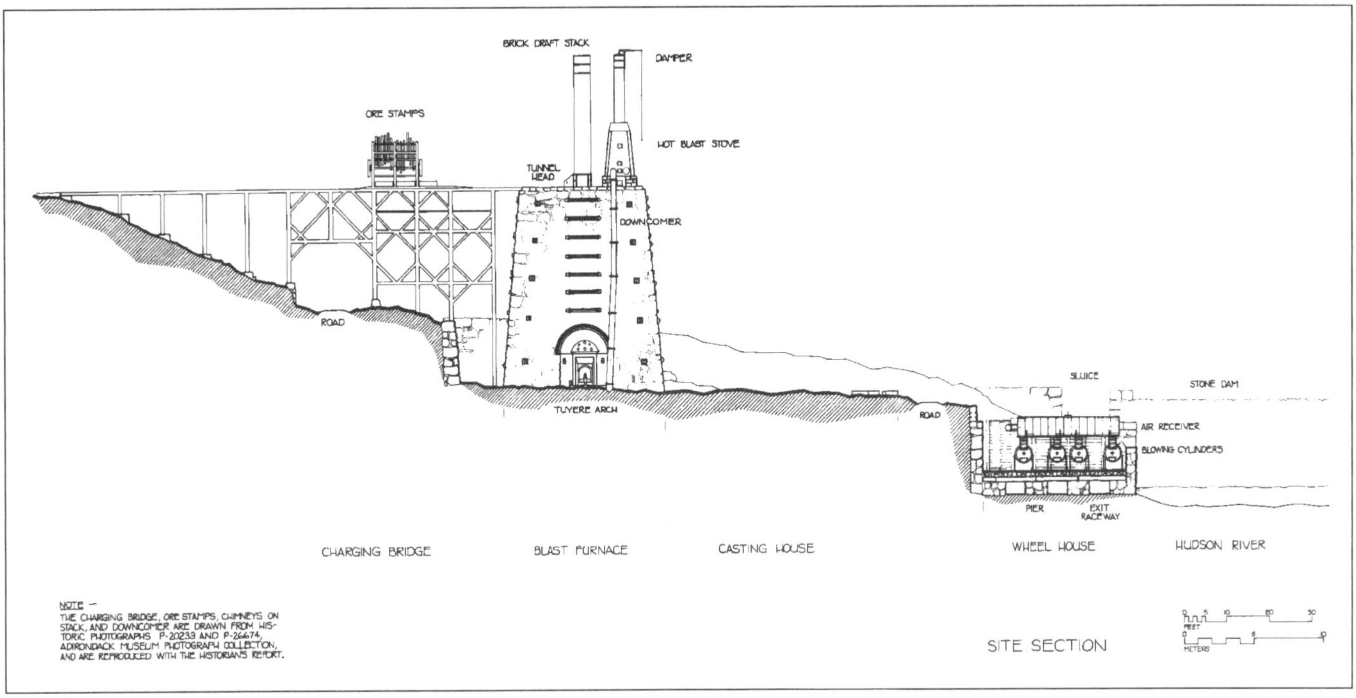
BRICK DRAFT STACK
DAMPER
ORE STAMPS
HOT BLAST STOVE
TUNNEL HEAD
DOWNCOMER
ROAD
TUYERE ARCH
ROAD
SLUICE
STONE DAM
AIR RECEIVER
BLOWING CYLINDERS
PIER
EXIT RACEWAY
CHARGING BRIDGE
BLAST FURNACE
CASTING HOUSE
WHEEL HOUSE
HUDSON RIVER
NOTE —
THE CHARGING BRIDGE, ORE STAMPS, CHIMNEYS ON STACK, AND DOWNCOMER ARE DRAWN FROM HISTORIC PHOTOGRAPHS P-20233 AND P-26674, ADIRONDACK MUSEUM PHOTOGRAPH COLLECTION, AND ARE REPRODUCED WITH THE HISTORIANS REPORT.
SITE SECTION
FEET
METERS

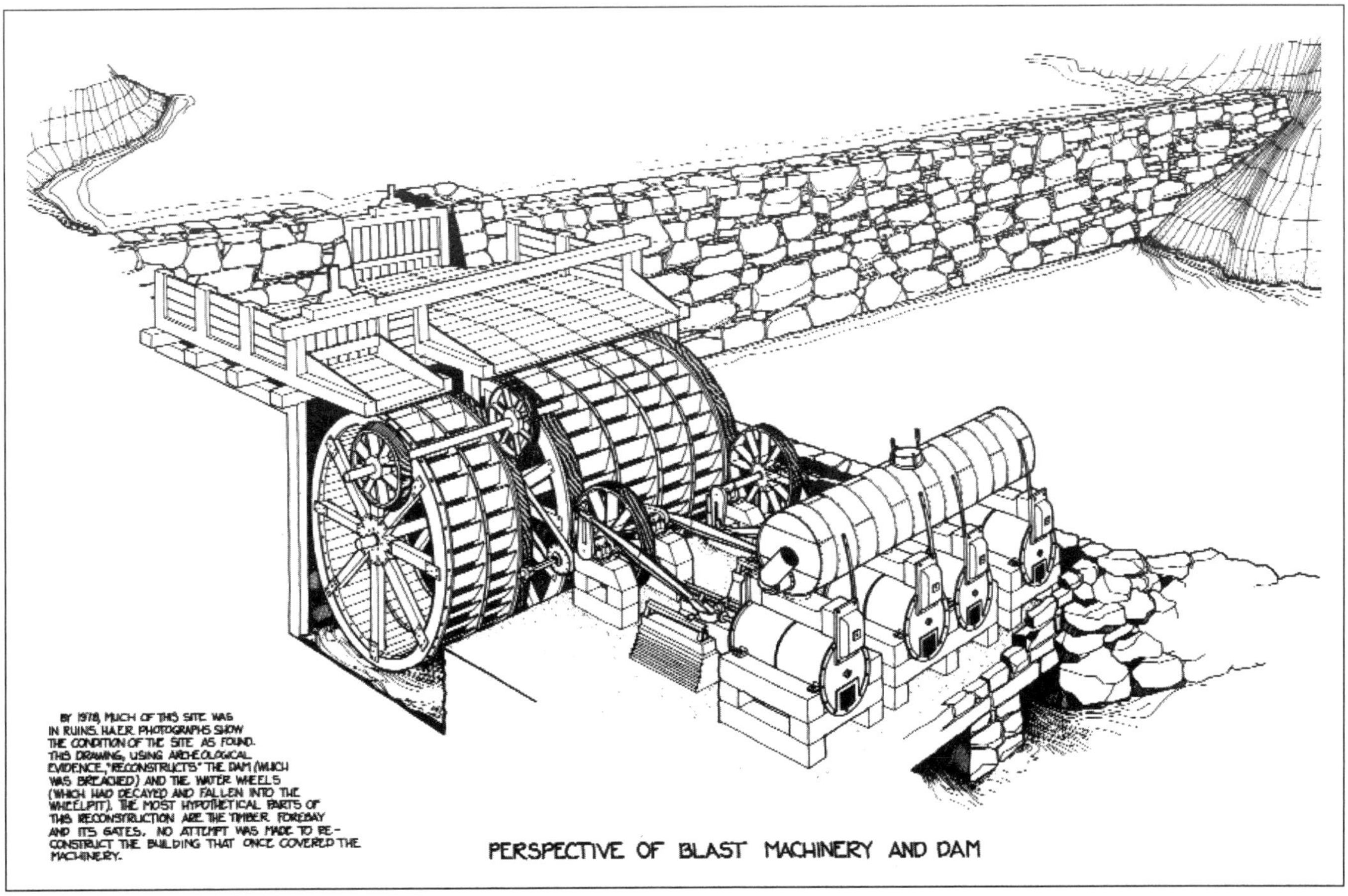
BY 1978, MUCH OF THIS SITE WAS IN RUINS. HAER PHOTOGRAPHS SHOW THE CONDITION OF THE SITE AS FOUND. THIS DRAWING, USING ARCHEOLOGICAL EVIDENCE, "RECONSTRUCTS" THE DAM (WHICH WAS BREACHED) AND THE WATER WHEELS (WHICH HAD DECAYED AND FALLEN INTO THE WHEELPIT). THE MOST HYPOTHETICAL PARTS OF THIS RECONSTRUCTION ARE THE TIMBER FOREBAY AND ITS GATES. NO ATTEMPT WAS MADE TO RE-CONSTRUCT THE BUILDING THAT ONCE COVERED THE MACHINERY.
PERSPECTIVE OF BLAST MACHINERY AND DAM

Tannehill Furnace, Bucksville, Alabama.

[LEFT] *Interior of one of the stacks as it appeared in 1936.*

[BELOW] *The furnace on the left was built in 1859, and the two furnaces on the right, in 1863.*

These three furnaces, between Birmingham and Tuscaloosa, Alabama, show a significant change in technology within only a few years' time. The first furnace was constructed in 1859 as a cold blast furnace. After the Civil War began, the Confederate government gave the owner of the furnace a contract to build two more furnaces. These he constructed in 1863 as hot blast furnaces, the blast machinery powered by a steam engine. The Tannehill furnaces are also significant because they were tended by slave labor, and there are archeological remains of slave quarters and a slave cemetery nearby. All three furnaces were partially destroyed by Union forces in 1865.

Sloss Furnaces, (No. 1 furnace on right, No. 2 on left), 1881-1882 (rebuilt 1927-1928), Birmingham, Alabama.

Less than 20 years after the last two Tannehill furnaces were constructed, a state-of-the-art iron-making facility was constructed in nearby Birmingham, Alabama. This furnace complex, the Sloss Furnace Company, was far different from Tannehill and represented the best of British and American furnace technology. The two original stacks at Sloss, built in 1881-1882, were twice the height of the Tannehill furnaces, had shells constructed of riveted iron plate instead of stone, used coke as fuel, and received their hot air blast from large free-standing stoves. But a combination of reliance on low-wage labor (instead of continually adopting labor-saving machinery) and the different quality of southern iron ores and coals as compared with ore from the Great Lakes region and Pennsylvania coal prevented Sloss and other southern iron and steel producers from following the path of northern producers. Unable to compete in the mass-produced iron and steel markets, Sloss found a niche in the iron-foundry business, which preferred traditional sand-cast pig iron, produced in small batches. Because of this congruence of technology, natural resources, labor practices, and market, Sloss held off modernizing until the 1920s, at which time the company adapted its furnace plant with new technology, equipment, and systems that had been in use in northern fields for more than 20 years.[13]

Blast Furnace Plant, Duquesne Works, Duquesne, Pennsylvania.

[ABOVE] *Ore bridges and ore yard (ore bridges date from 1954 and 1962), in front of a row of blast furnaces.*

[FACING PAGE] *Isometric of Blast Furnace No. 1 (originally constructed in 1896).*

The iron industry reached its zenith of nineteenth-century technological development in Duquesne, Pennsylvania, in what became known as the "Duquesne Revolution." The Duquesne Works had been built in the late 1880s and acquired by Andrew Carnegie in 1890. A new blast furnace plant constructed between 1895 and 1897 featured the larger blast furnaces and more powerful blowing engines that Carnegie had been developing in his various works since the 1870s, but it also included innovative advances in the delivery of raw materials to the furnace stacks. The industry's first automatic raw materials storage, handling, and delivery system was invented at Duquesne and included coordinated ore-storage facilities, stocking bridges, hoisting apparatus, and furnace-charging facilities. This innovation increased production, drastically reduced labor costs, and made the Duquesne furnace plant the most modern in the world.

THIS DRAWING REPRESENTS THE RAW MATERIALS DELIVERY, SMELTING, AND SLAG CASTING SYSTEMS AT BLAST FURNACE NO. 1. SMELTING TOOK PLACE AFTER THE FURNACE HAD BEEN STOCKED WITH COKE, IRON ORE, AND LIMESTONE AT A WEIGHT RATIO OF 4:2:1. AT THE TOP OF THE FURNACE, THE STOCK WAS DRIED BY COMING INTO CONTACT WITH AN ASCENDING CURRENT OF CARBON MONOXIDE AND OTHER HOT GASES PRODUCED BY THE CONTACT OF HOT BLAST AIR WITH COKE AT THE LEVEL OF THE FURNACE TUYERES. AS THE STOCK DESCENDED IN THE FURNACE, THE CARBON MONOXIDE OXIDIZED THE ORE INTO A SPONGY, FREE IRON STATE AND LIQUIFIED THE LIMESTONE. AT THE FURNACE FUSION ZONE OR BOSH, THE IRON LIQUIFIED AND LIQUID LIME COMBINED WITH THE MINERALS WHICH HAD BEEN SEPARATED FROM THE ORE, FORMING SLAG. FROM THIS POINT BOTH THE SLAG AND THE MOLTEN IRON TRICKLED DOWN PAST THE COMBUSTION ZONE OF THE FURNACE TO ITS HEARTH THROUGH THE INTERSTICES OF THE COKE. DUE TO ITS LIGHTER WEIGHT, THE SLAG FORMED A MOLTEN LAYER ON TOP OF THE IRON.

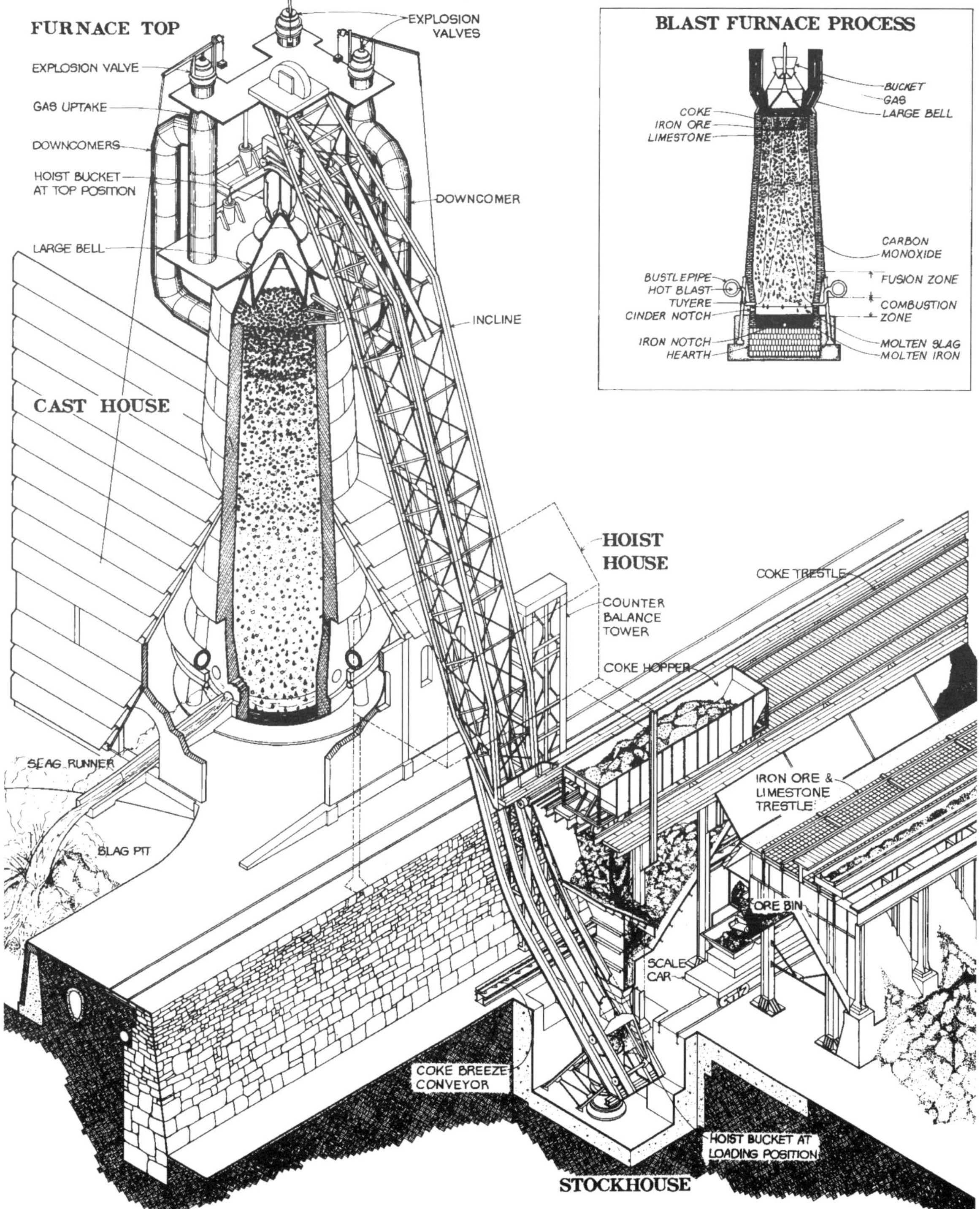

ISOMETRIC OF BLAST FURNACE NO.1

REFERENCE DRAWINGS

CSC DWG NOS: 13090, 14911, 14964, 14970, 15219, 15322.
USS DWG NOS: 25998, 27040.
AB DWG NOS: E-4791 #1,2,4,5; E-4792 #5;
E-4795 #7,E1; E-4796 #1,2,E3; E-5574 #1-4,E1,E2.

SCALE 3/32"=

With Heat and Hammer: From Charcoal to Carriages

Kiln No. 2, Frisco Charcoal Kilns, 1877, near Milford, Utah.

Kilns such as these were used throughout the West in the late nineteenth century to produce charcoal to fuel smelters in western mining districts. Coke, a better smelting fuel, eventually replaced charcoal when transportation rates and production costs for coke decreased.

[LEFT] *Minett Varnish Kiln, 1853, Clifton, New Jersey.*

[BELOW] *Kiln, Salt Lake Pressed Brick Company, 1891, Salt Lake City, Utah.*

These kilns were used to dry and bake their products: varnish and bricks. These industries were difficult to completely mechanize in the nineteenth century, as intensive hand labor and worker knowledge both were required in the production process. A writer for *Scientific American* commented on brick making in 1886: "No rule can be laid down for the handling of clay; the routine which in one yard produces first quality would, if transferred without change to another, only cause miserable failure."[14]

Griffin No. 1 Coke Works, Bessemer Coke Company, c. 1900, near Masontown, Pennsylvania.

The Bessemer Coke Company opened its Griffin No. 1 Coal and Coke Works in Fayette County in Pennsylvania sometime after 1899. Part of the works included a battery of double-block rectangular coke ovens, an improved version of the old beehive coke ovens. When in full production, Griffin No. 1 had 310 ovens in operation, producing approximately 150,000 tons of coke per year.

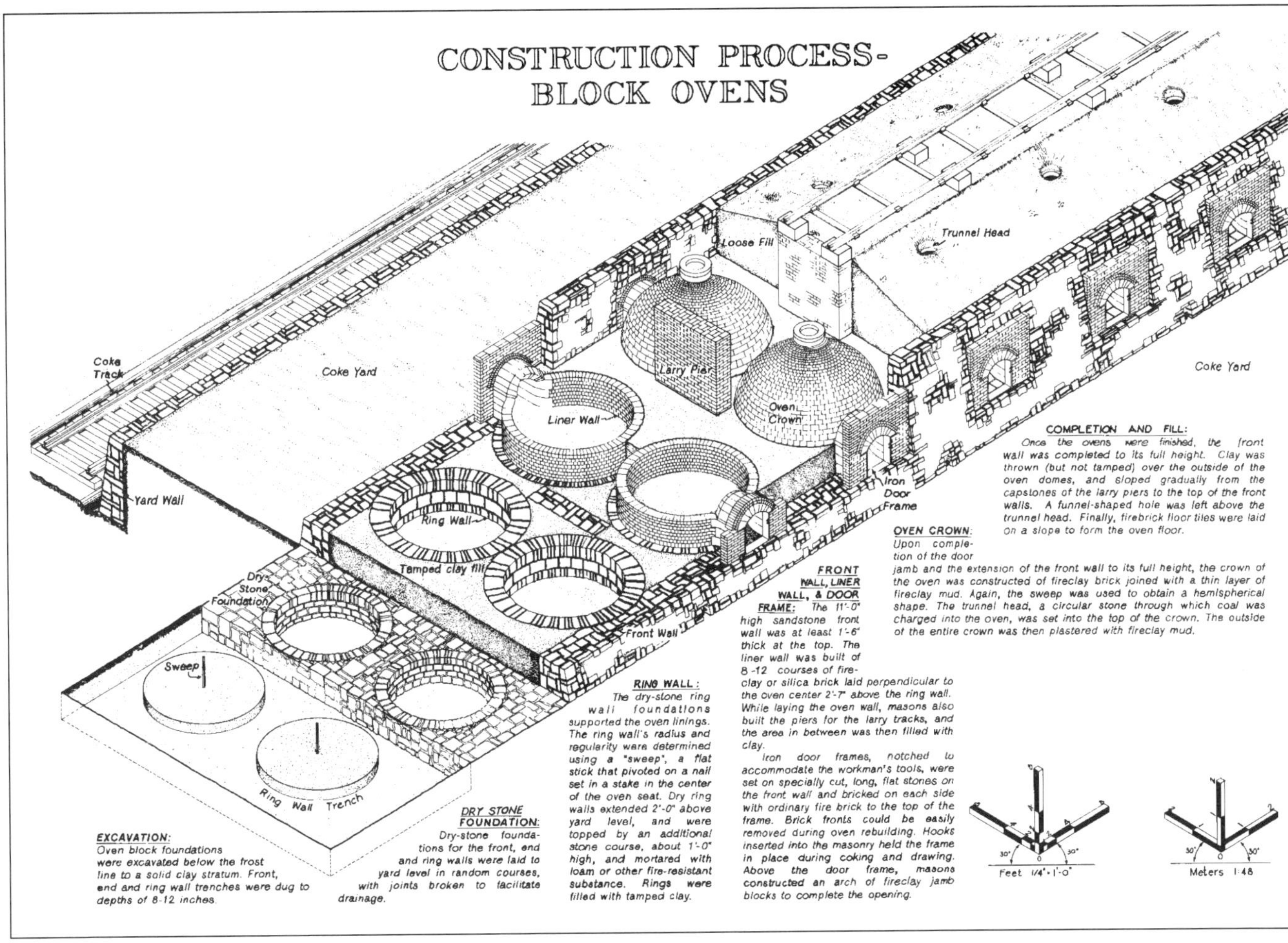

Block oven construction process, near Connellsville, Pennsylvania.

Birkenhead Powder Mill (with reconstructed waterwheel), Du Pont Powder Mills, near Greenville, Delaware.

In 1802, Irenee du Pont, a recent émigré from France, constructed a black-powder mill along the banks of the Brandywine River in Delaware. Black powder was a mixture of saltpeter, sulfur, and charcoal, and after these elements were refined, they were mixed together in rolling mills. The mills took their form because of the explosive nature of black powder. Cast-iron rolling wheels were surrounded by three heavy stone walls, a light roof, and a light fourth wall that faced the stream and directed any accidental blast over the water and away from the rest of the mill complex and community. The mills were first powered by waterwheels and later by water turbines.

City Blacksmith Shop, c. 1898, Lamberton, Minnesota.

Blacksmiths were an integral part of early American society, providing horseshoes, plow and wagon parts, farming implements, household wares, tools, building hardware, and other necessities. Blacksmithing had its repetitious moments, but perhaps more than most other craftsmen, blacksmiths were versatile fabricators. In one Massachusetts smith's 1842 account book, for example, he recorded making or repairing such varied items as sleds, wagons, plows, hooks, chains, a dung fork, hammers, wheelbarrows, a pitch fork, hinges, an axe, buckets, bolts, hoops, tongs, drills, hoes, rings, kettles, and a boat. Blacksmiths were not always located in isolated villages under spreading chestnut trees, however. Almost all large nineteenth-century industrial complexes contained a blacksmith's shop, a vital necessity for producing forged parts and keeping machinery functioning. But as the iron and steel industry matured in the nineteenth century, and as automobiles eventually replaced horses in the next, blacksmith's shops were relegated to antiquity, to be replaced by machine shops and foundries. [15]

Gruber Wagon Works, 1883, near Bernville, Pennsylvania.

[FACING PAGE]

[TOP] *Gruber Wagon Works, c. 1895.*

[BOTTOM] *Tire setter and press in the blacksmith shop.*

[THIS PAGE]

[RIGHT] *Spoke-tenoning machine in the wood shop.*

[BELOW] *A Gruber-built hay flat and wagon.*

The Gruber family of Berks County, Pennsylvania, began operating a wagon works as early as 1854. Franklin H. Gruber built a new shop in 1883 that produced farm wagons for the local region. Powered by a water turbine, the shop combined hand labor and small machinery to mass-produce wagons. Production peaked in the 1920s, when the shop turned out an average of 100 wagons a week. The entire structure was relocated 5 miles in 1976 to make way for a dam and lake.

[RIGHT] *Seneca Glass Company, 1896, Morgantown, West Virginia.*

The Seneca Glass Company was started by immigrant German glassblowers in 1896. In an industry undergoing rapid mechanization, Seneca glassblowers concentrated on specialty items such as leaded-glass tableware, thus preserving a semblance of the traditional craft process. This tradition was wedded to new technology, however, such as the use of natural gas as a fuel, which allowed for a larger melting furnace.

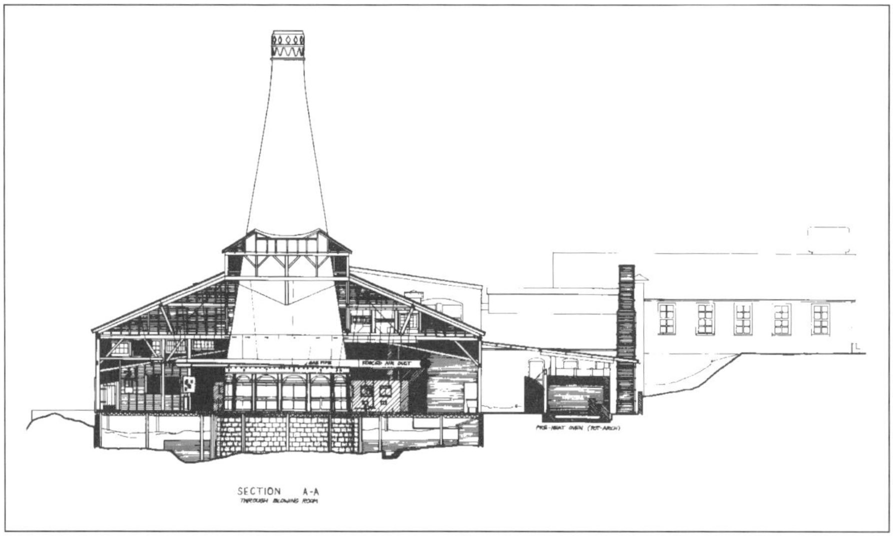

[BELOW] *Kilns, Coplay Cement Company, 1871, Coplay, Pennsylvania.*

[FACING PAGE] *Kilns, Alamo Roman and Portland Cement Company, 1880, San Antonio, Texas.*

Concrete was more widely adopted as a building material after 1871 when David Saylor began to manufacture Portland cement, one of the ingredients of modern concrete, in Coplay, Pennsylvania. Saylor's material was endorsed by the federal government when his cement was specified for James B. Eads's South Pass jetties at the delta of the Mississippi River in 1875.[16] The Alamo Roman and Portland Cement Company began to manufacture Portland cement in San Antonio, Texas, in 1880 and modernized its production in 1897 with a rotary kiln, which produced cement at a faster pace.

A CLOSER LOOK

By the Sweat of Their Brow: Workers

"It was a pleasure to see him handle a tool of any kind, but he was quite splendid with an eighteen inch file."[17]

American workers were extraordinarily productive in the nineteenth century, and many technological innovations and advancements were achieved by workers on the shop floor as they figured out adjustments and reconfigurations of machinery and processes.[18] Whether American industrialization was ultimately in the best interest of workers has been debated for decades, however. The growth and expansion of industry gave work to millions and often brought an improvement in living conditions, but the regimentation, monotony, and dangers of factory and millwork led to massive and violent protest by the century's end. Still, many workers took enormous pride in the engineering and technological works they helped construct in the nineteenth century, and when a photographer asked them to pose for a photograph; they tended to pose as conquering heroes, builders of a new world.[19]

[ABOVE] *Shop crew in the Chicago, Burlington and Quincy Railroad Roundhouse, photographed c. 1878-1885, Aurora, Illinois.*

A CLOSER LOOK

By the Sweat of Their Brow: Workers

[RIGHT] *Clock tower, Boott Cotton Mills, Lowell, Massachusetts.*

[BELOW LEFT] *Front gate, Boott Cotton Mills.*

[BELOW RIGHT] *Interior of privy tower, Hopkins Mill, West Greenwich, Rhode Island.*

Bells, clocks, gates, privy towers—as industry and engineering became more rationalized and time-conscious in the nineteenth century, the physical environment for workers reflected this ethos. Bells and clocks regulated work time, gates demarcated workspace from home and community, and even privies were judiciously placed to minimize workers' time away from their stations.

A CLOSER LOOK

By the Sweat of Their Brow: Workers

[RIGHT] *Work crew on the Forty-Sixth Street Bridge, 1895, Ashtabula, Ohio.*

[BELOW] *Workers in the Erecting Shop of the P.H. and F.M. Roots Company Factory, makers of rotary blowers, c. 1890, Connersville, Indiana.*

A CLOSER LOOK

By the Sweat of Their Brow: Workers

Cannelton Cotton Mill, Cannelton, Indiana.

[ABOVE] *Workers and their families in front of workers' houses (unknown date).*

[LEFT] *Workers' housing, built c. 1855.*

When Cannelton Cotton Mills opened in southern Indiana in 1851, many of the workers were young women from New England who were brought west under 2-year contracts. Replicating the style in New England, houses and a hotel were built to accommodate the new labor force.

A CLOSER LOOK

By the Sweat of Their Brow: Workers

Breaker building and church, Avery Island Salt Works, Avery Island, Louisiana.

A variety of factors, including site isolation and company paternalism, led many nineteenth-century industries to establish company towns for their workers, complete with company-owned churches, stores, and houses. From the Lowell boarding houses of the first half of the century to mine and lumber towns of the latter half, companies constantly experimented with methods to attract, retain, and control workers.

How Things Work: Power

[LEFT] *Burden Water Wheel, built in 1851, Troy, New York.*

The "Niagara of Water Wheels" was built in 1851 by ironmaster Henry Burden to power his horseshoe factory and ironworks in Troy, New York. The mammoth wheel, which had a diameter of 60 feet, was one of the most powerful waterwheels ever built. It was destroyed in 1942 in a scrap iron drive.[20]

[RIGHT] *Power schematic, Mount Pleasant Grist Mill, c. 1805, near St. Peters, Pennsylvania.*

Water provided much of the power to operate American machinery in the nineteenth century, especially in the first half of the century. Waterwheels, such as this one for a grist mill, and water turbines turned a complex set of gears and belting to perform a variety of tasks.

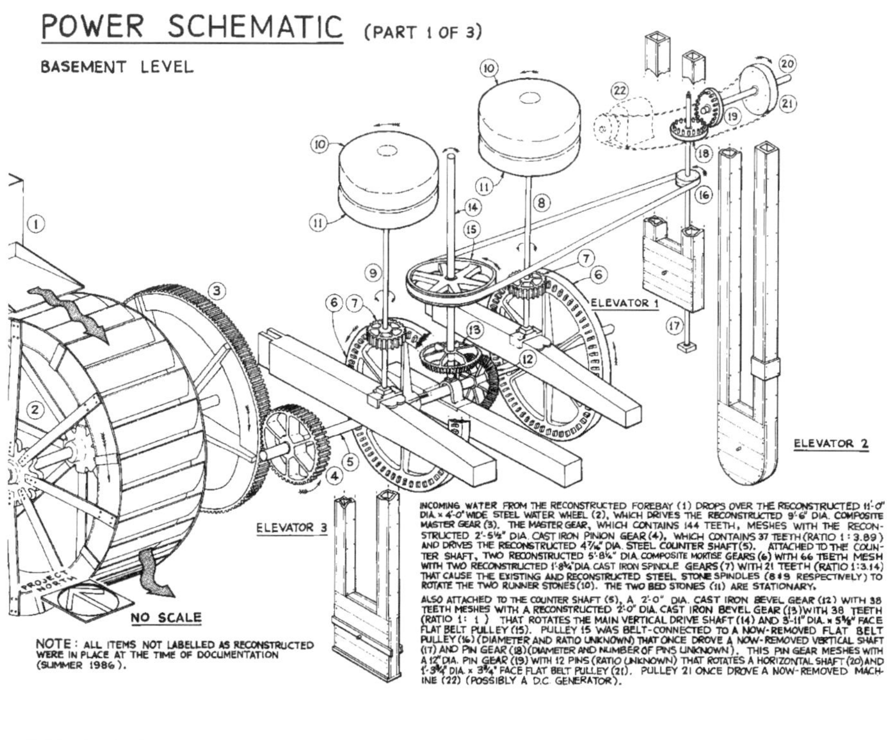

[RIGHT] *Belting transmitting power to the mill's machinery from water turbines, Harmony Manufacturing Company, Mill No. 3, 1866-1872, Cohoes, New York.*

[LEFT] *Sugar mill steam engine, 1861, Hacienda Azucarera La Esperanza Sugar Mill, Manati, Puerto Rico.*

During the second half of the nineteenth century, steam became the dominant power source. The West Point Foundry of Cold Spring, New York, manufactured in 1861 both this Gothic-style beam engine and cane mill, which crushed sugar cane. The La Esperanza engine is the only West Point Foundry beam engine known to survive and the only known six-column beam engine made by any American manufacturer.

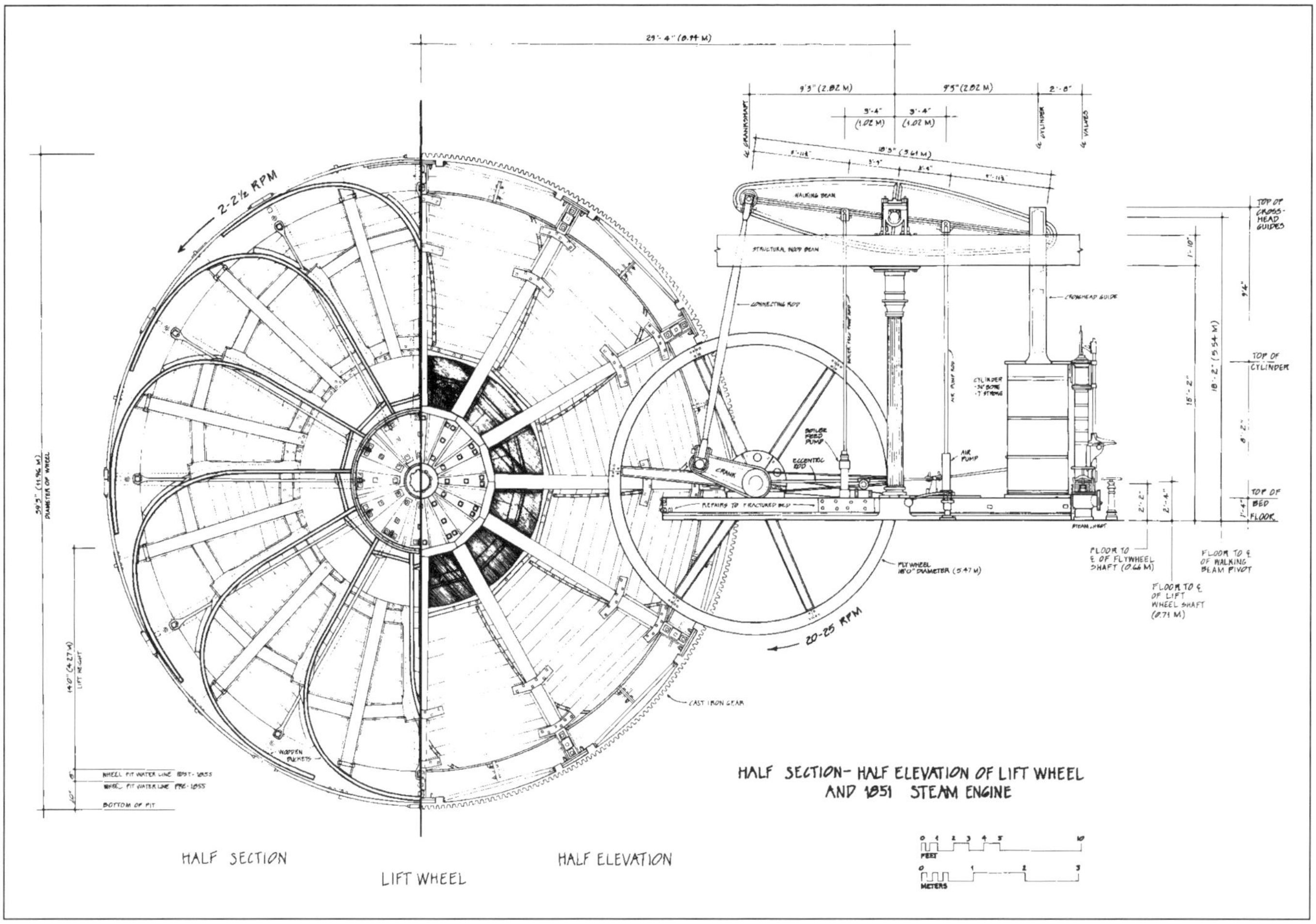

Section and elevation of lift wheel and steam engine, 1851, Chesapeake and Delaware Canal Pump House, Chesapeake City, Maryland.

The Chesapeake and Delaware Canal opened in 1829 to connect the Chesapeake Bay and the Delaware River. From the beginning, however, there was not enough water along the summit level of the canal to make up for that lost in lockage, and in 1851 Samuel Merrick and John Towne of the firm of Merrick and Sons of Philadelphia, Pennsylvania, installed a 39-foot diameter lift (or scoop) wheel and a system of raceways that transferred water from a nearby creek into the canal. The wheel, powered by a steam engine, delivered more than 200,000 cubic feet of water per hour into the canal. A second engine was added in 1854, and the lift wheel was replaced in 1856, but no other changes were made in the system until 1927, when the canal was renovated as a sea-level canal without locks.

Dorrance Colliery Fan Complex, 1883, Wilkes-Barre, Pennsylvania.

[RIGHT] *Steam engine for mine fan.*

[BELOW] *Steam engine and Guibal fan.*

Anthracite coal was mined at the Dorrance Colliery near Wilkes-Barre, Pennsylvania, beginning in 1883. Ventilating fans were necessary to provide fresh air to the underground workings and to remove noxious gases. This steam engine and Guibal fan were manufactured in 1883 by the Pittston Engine and Machine Company of West Pittston, Pennsylvania. The fan, of wood and cast iron, was 35 feet in diameter and was driven by the steam engine at a rate of 49 revolutions per minute. When in active service, it ran 24 hours a day.

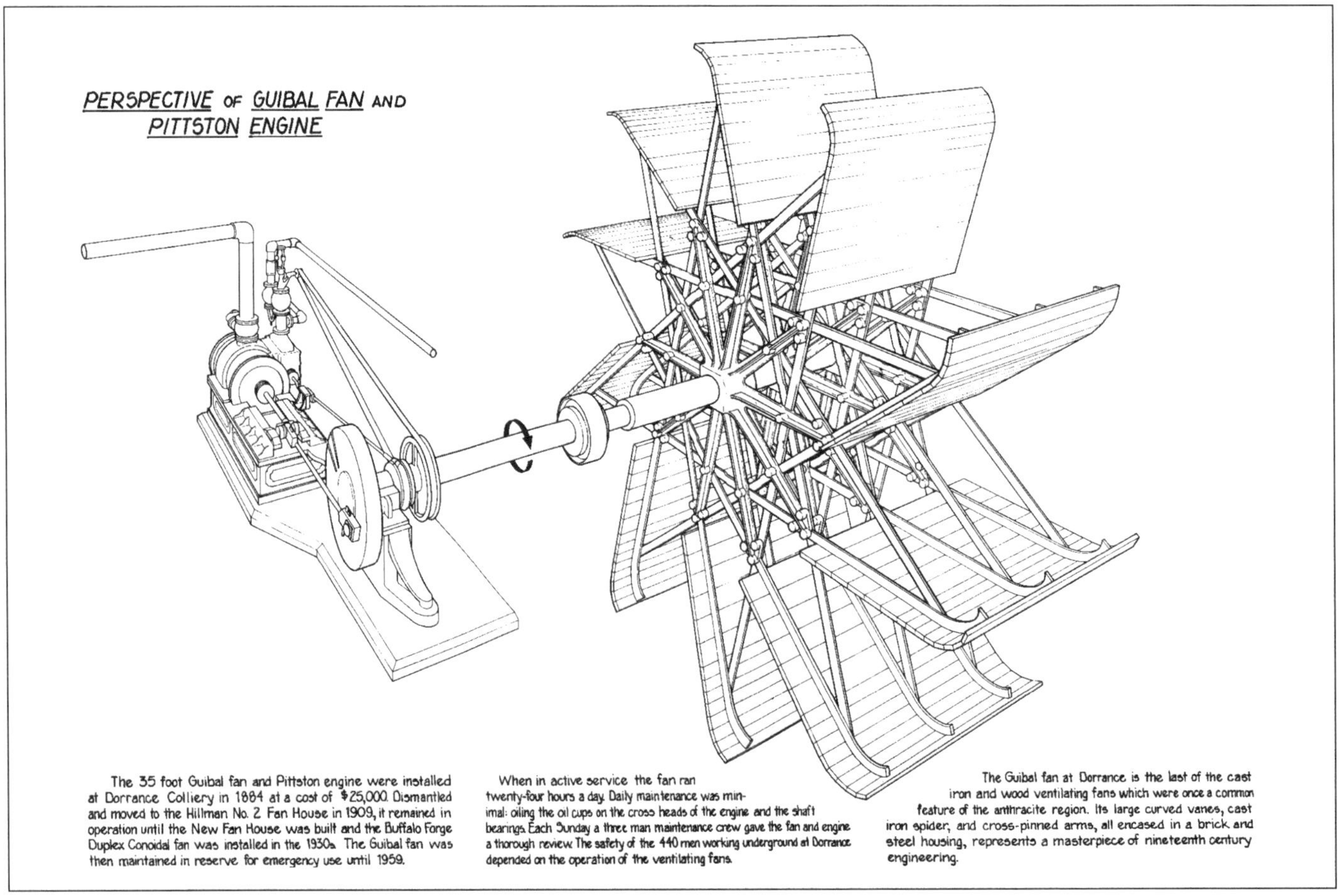

Line shaft, pulleys, and belts of the hoisting machinery, Pawtucket Gatehouse, Northern Canal, Lowell, Massachusetts.

Power was transmitted throughout much of the nineteenth century by leather belt drives, which were gradually replaced in the next century by electric motors.

Machine Shop, East Broad Top Railroad, Rockhill Furnace, Pennsylvania.

[RIGHT] *1882 steam engine in the Machine Shop powered the overhead lineshaft system in the railroad shops.*

[BELOW] *Overhead lineshafting in Machine Shop.*

A complex arrangement of overhead lineshafts in the East Broad Top Railroad Machine Shop was powered by a steam engine, erected in 1882. The system included nearly 36 separate shafts that powered machinery in several of East Broad Top's shops.

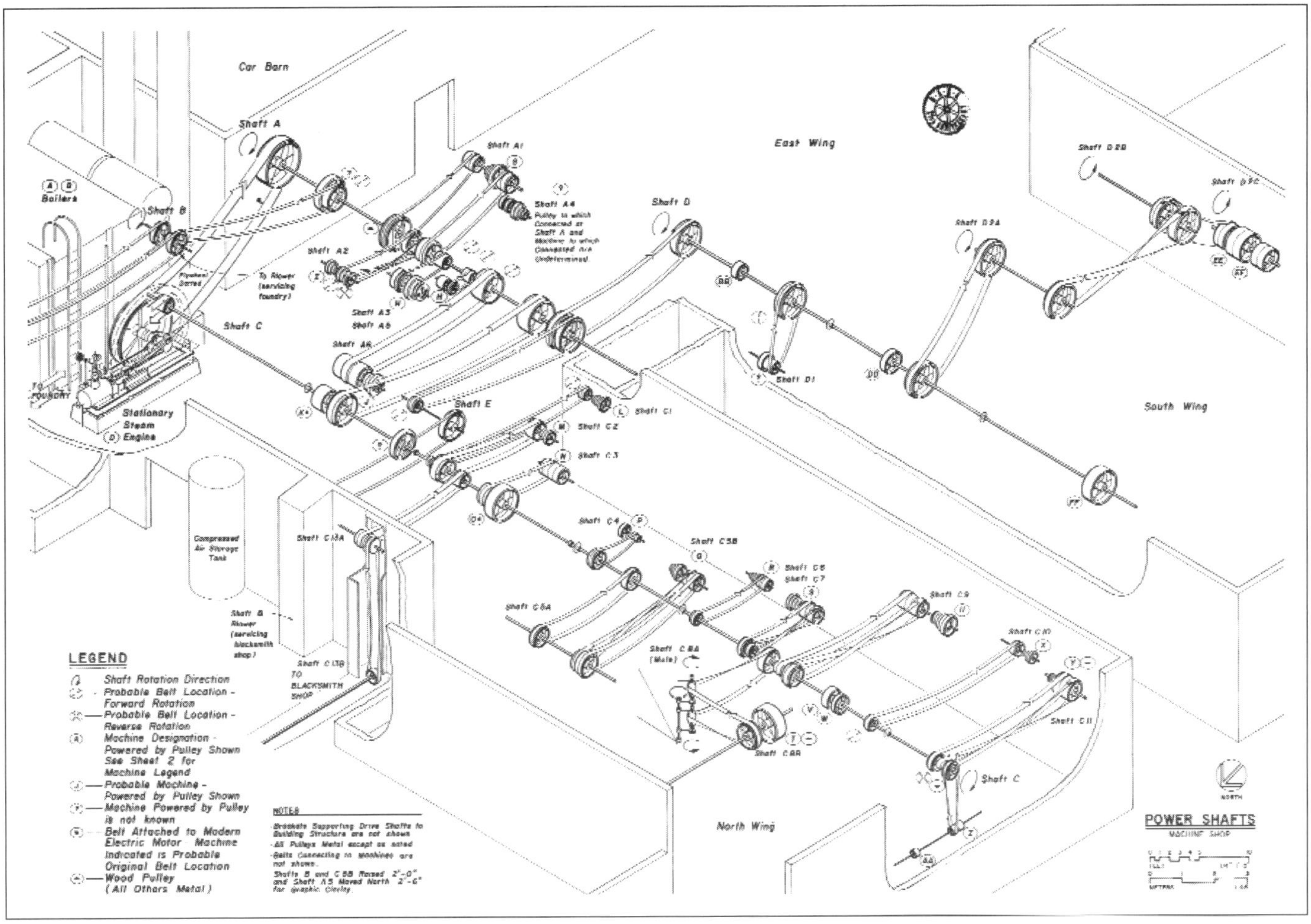

[ABOVE] *Lineshafting in the Machine Shop.*

[LEFT] *Leavitt Pumping Engine, 1894, Boston Water Works, Boston, Massachusetts.*

The Leavitt Pumping Engine was installed in the Chestnut Hill High Service Pumping Station of the Boston Water Works in 1894. Designed by Erasmus D. Leavitt, Jr., a prominent American mechanical engineer, this triple-expansion steam pumping engine was capable of pumping 20 million gallons of water per day.

Folsom Powerhouse, 1895, Folsom, California.

[ABOVE] *Interior.*

[LEFT] *Generator No. 4.*

[FACING PAGE]
Governor of Turbine No. 3.

The Sacramento Electric Power and Light Company constructed the Folsom Dam and Powerhouse on the American River in California in the early 1890s. Folsom was not the first hydroelectric power plant in the country, but when in 1895 it successfully transmitted power to Sacramento, California, 24 miles away, it achieved the longest commercial power transmission to date. Equipment in the powerhouse included four water turbines driving 750-kilowatt General Electric generators.[21]

Power plant, 1900, Pratt Institute, Brooklyn, New York.

When the Pratt Institute, a technical training school in Brooklyn, New York, expanded in the late 1890s, new electrical equipment was added to the power plant. This included three new steam engine generators manufactured by General Electric and single-cylinder engines by Ames Iron Works of Oswego, New York.[22] These were among the oldest steam-powered electric generators in the country when recorded by the Historic American Engineering Record in 1974.

References

1. Greville Bathe and Dorothy Bathe, *Oliver Evans: A Chronicle of Early American Engineering* (Philadelphia Historical Society of Pennsylvania, 1935) pp. 190–206; quotation from p.191.

2. Siegfried Giedion, *Mechanization Takes Command* (New York: Norton, 1969 [originally published in 1948]) pp. 79–86; quotations from p. 85.

3. Brooke Hindle and Steven Lubar, *Engines of Change: The American Industrial Revolution, 1790–1860* (Washington, D.C.: Smithsonian Institution Press, 1986) p. 158; Russell Blaine Nye, *Society and Culture in America, 1830-1860* (New York: Harper, 1974) p. 272.

4. Hindle and Lubar, *Engines of Change*, pp. 158–160.

5. Quotation by Salmon P. Chase, as quoted in Nye, *Society and Culture in America, 1830–1860*, p. 272.

6. Louis C. Hunter, *A History of Industrial Power in the United States, Vol. II: Steam Power,* (Charlottesville, VA: University Press of Virginia, 1985) p. 432.

7. Theodore Z. Penn, "The Development of the Leather Belt Main Drive," *IA, Journal of the Society for Industrial Archeology*, 7 (1981):1-14.

8. Louis C. Hunter and Lynwood Bryant, *A History of Industrial Power in the United States, Vol. III: The Transmission of Power* (Cambridge, MA: MIT Press, 1991) p. 119.

9. Hunter, *A History of Industrial Power in the United States, Vol. II: Steam Power*, pp. 433–434.

10. Hindle and Lubar, *Engines of Change*, p. 192.

11. David E. Nye, *American Technological Sublime* (Cambridge, MA: MIT Press, 1994) p. 115.

12. Bruce E. Seely, "Blast Furnace Technology in the Mid-19th Century: A Case Study of the Adirondack Iron and Steel Company," *IA, Journal of the Society of Industrial Archeology* 7 (1981):47.

13. W. David Lewis, *Sloss Furnaces and the Rise of the Birmingham District: An Industrial Epic* (Tuscaloosa, AL: University of Alabama Press, 1994) pp. 74–79, 98–99, 391–411, 474–504.

14. Kim E. Wallace, *Brickyard Towns: A History of Refractories Industry Communities in South-Central Pennsylvania* (Washington, D.C.: National Park Service, 1993) p. 22.

15. Aldren A. Watson, *The Village Blacksmith* (New York: Thomas Y. Crowell, 1968) pp. 112–116.

16. Carl W. Condit, *American Building Art: The Nineteenth Century* (New York: Oxford University Press, 1960) pp. 227–228.

17. A fellow workman's description of a master mechanic, as quoted in Lewis Mumford, *Technics and Civilization* (New York: Harcourt, Brace, 1934) p. 210.

18. Patrick M. Malone, "Little Kinks and Devices at Springfield Armory, 1892–1918," *IA, Journal of the Society of Industrial Archeology* 14 (1988):59–76.

19. F. Jack Hurley, ed., *Industry and the Photographic Image* (New York: Dover, 1980) pp. 2–3.

20. Robert Vogel, "The Burden Water-Wheel," *Occasional Publications, No. 2,* (Society for Industrial Archeology, 1973).

21. American Society of Mechanical Engineers International History and Heritage Committee, *Landmarks in Mechanical Engineering* (West Lafayette, IN: Purdue University Press, 1997) pp. 62–63.

22. American Society of Mechanical Engineers International History and Heritage Committee, *Landmarks in Mechanical Engineering*, pp. 82–83.

3

Structures and Materials

The nineteenth century was an age of great experimentation in building forms, materials, and techniques. According to Carl Condit, the "explosive progress of building techniques produced in one century more innovations in material, structure, form, and method of construction than the whole previous history of the art."[1] Wood was the dominant building material through the first half of the century because of America's vast forests, the high labor costs associated with iron and stone fabrication, and the American tendency to value speed of erection over permanency. Stone was used for structures as well, but it was too expensive to quarry and shape for most builders. Because of its cheapness and load-bearing capacity, brick was a popular building material for industrial structures throughout the century. Following the English example, American engineers experimented with iron construction in the 1830s and 1840s. The first iron bridge appeared in 1838, followed by an iron-framed building a decade later.[2] Structural wrought-iron beams were developed in the 1850s, providing the forms and techniques later used with steel beams. Toward the end of the century, again following English and European precedent, concrete was used as a building material, but the widespread use of reinforced concrete in bridge and factory construction did not occur until the twentieth century.

By the end of the nineteenth century, iron and steel had replaced wood, stone, and brick as the primary building materials for engineers, and reinforced concrete was just beginning to emerge as a structural material. But this transition (or transitions) did not occur in a neat and orderly fashion; there was much experimentation (and a few disasters) before particular structural forms were adopted, and older forms and materials were still in use at century's end. During the middle decades of the century, for example, wood, cast iron, and wrought iron often were used in combination in trusses for bridges and industrial buildings as engineers sought the best configurations within parameters of structural performance, material availability, cost of fabrication, building use, and client desires. Wood and brick, to take another example, remained popular building materials even in the late nineteenth century because of the materials' cheapness, beauty, and ease of fabrication.

Engineering (and technology in general) had become more standardized, scientific, and mechanized by 1900, and new techniques and materials led to ever larger and longer structures. But the inventor-mechanic tradition lingered in America throughout the century, leading to the creation of "vernacular" technology that at times defied tradition and the norms of the engineering profession. Pipe bridges, hand-dug canals, and harrowing cliff-side roads are testaments to the imagination, inventiveness, and persistence of the American people in the nineteenth century.

[FACING PAGE] *Builders' plate of the Hale Bridge, a bowstring truss bridge near Oxford Junction, Iowa, built in 1877-79.*

[INSET] *Numbered rafters, Slater Mill, Pawtucket, Rhode Island.*

1879
BUILT BY
NG IRON BRIDGE Co
CLEVELAND
O

The Structure of Structures: Materials and Connections

[ABOVE] *Interior, Mills No. 1 and No. 2, 1865, Boott Cotton Mills, Lowell, Massachusetts.*

[RIGHT] *Wooden columns and beams, Pier G (a covered freight pier), 1891, Lehigh Valley Railroad, Jersey City, New Jersey.*

Wood was the dominant building material at the beginning of the nineteenth century, and although the use of iron, for example, created more space in mills and factories, cheaper heavy-timber framing remained a viable alternative throughout the century.

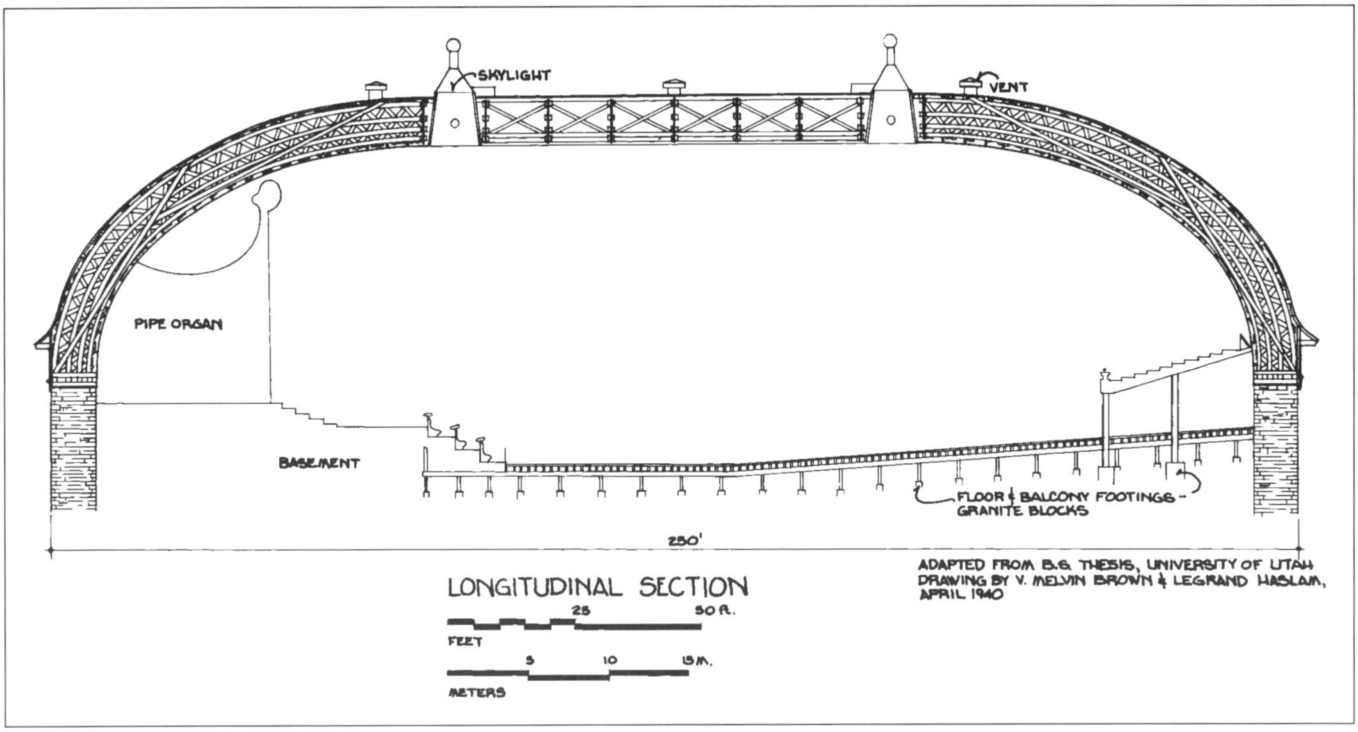

Salt Lake City Tabernacle, 1864 and 1867, Salt Lake City, Utah.

[ABOVE] *Elevation of a roof truss.*

[RIGHT] *Part of a timber arch.*

The Salt Lake City Tabernacle of the Church of Jesus Christ of the Latter Day Saints has been called the "greatest work of timber framing still surviving" in America. A massive structure measuring 150 feet by 250 feet, the Tabernacle is composed of a complex series of wooden trusses, half domes, vaults, and bracing. Much of the credit for its design belongs to Henry Grow (1817-1891), a bridge builder from eastern Pennsylvania who migrated west to Utah as a convert to the Mormon church. The roof of the Tabernacle is made up of a central vaulted area flanked by half domes. All of the timber connections are mortised, pinned, and wedged for tightness. The building of the Tabernacle is even more extraordinary considering that Salt Lake City, Utah, in the early 1860s was 1,500 miles from the nearest railroad and that all elements of the structure thus had to be obtained and made locally.[3]

[ABOVE] *Stone arch, Jefferson Street Bridge, 1857, Jefferson, Missouri.*

[LEFT] *Arched drain opening, Brick Arch Viaduct, Central of Georgia Railway, 1853, Savannah, Georgia.*

Stone and brick construction was used throughout the century for technological structures. Durable, fire-resistant, and familiar, masonry served a wide variety of building purposes. Although the use of stone was somewhat limited because of high transportation and labor costs, brick was popular for industrial buildings and later used in combination with iron and steel.

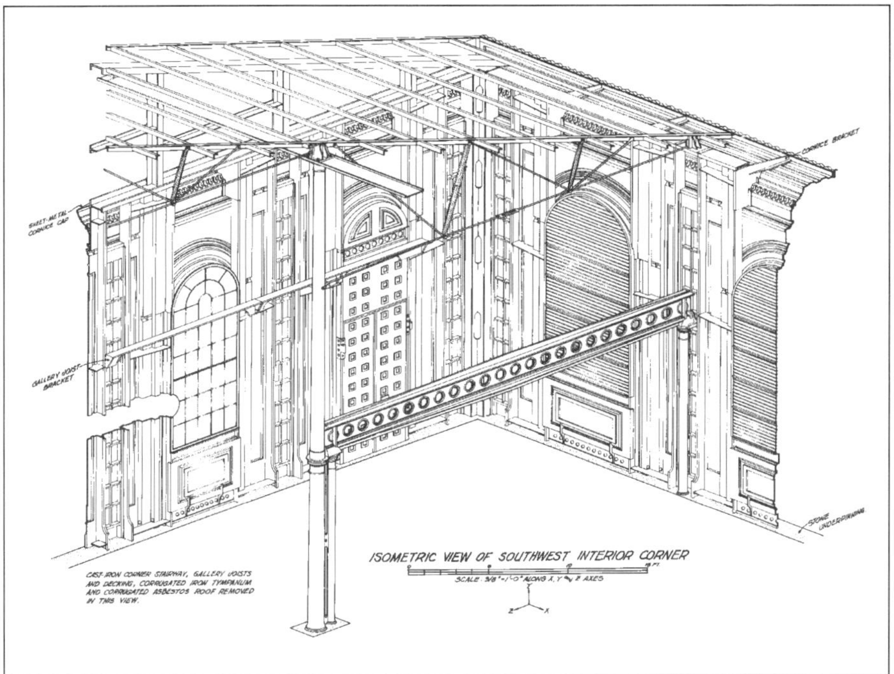

Iron Storehouse, 1859, Watervliet Arsenal, Watervliet, New York.

[ABOVE] *Iron framing of the storehouse.*

[RIGHT] *Iron columns supporting the gallery.*

[BELOW] *Elevation of the storehouse.*

In an era of experimentation in metal fabrication, this storehouse for the Watervliet Arsenal, completed in 1859, is an outstanding example of all-metal construction. Composed almost entirely of cast and wrought iron, the structure's elements were prefabricated by the Architectural Iron Works of New York, shipped up the Hudson River, and erected on site.

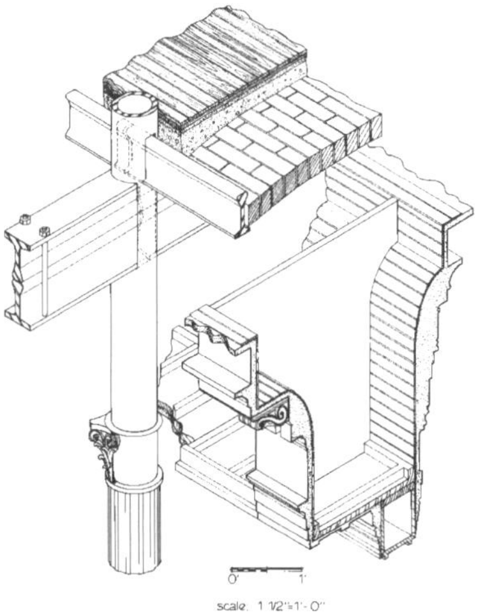

isometric projection of structure and molding - galleria ceiling , 1859

Cooper Union for the Advancement of Science and Art, 1853-1859, New York City, New York.

[FACING PAGE]

[TOP] *Exterior.*

[BOTTOM LEFT] *Beam and column construction.*

[BOTTOM RIGHT] *Exposed ends of wrought-iron structural beams on brick arches.*

[THIS PAGE]

[LEFT] *Ventilation holes under Great Hall seating.*

[BOTTOM] *Fan and ventilation system of the Great Hall, as improved in 1888.*

The Cooper Union, completed in 1859, was technologically innovative in addition to having a pioneering role in free public education. The structure was among the first buildings in the United States to use structural wrought-iron beams, and it also contained innovative ventilation and plumbing systems, as well as a circular shaft in anticipation of an elevator, an invention that had not yet been perfected at the time.

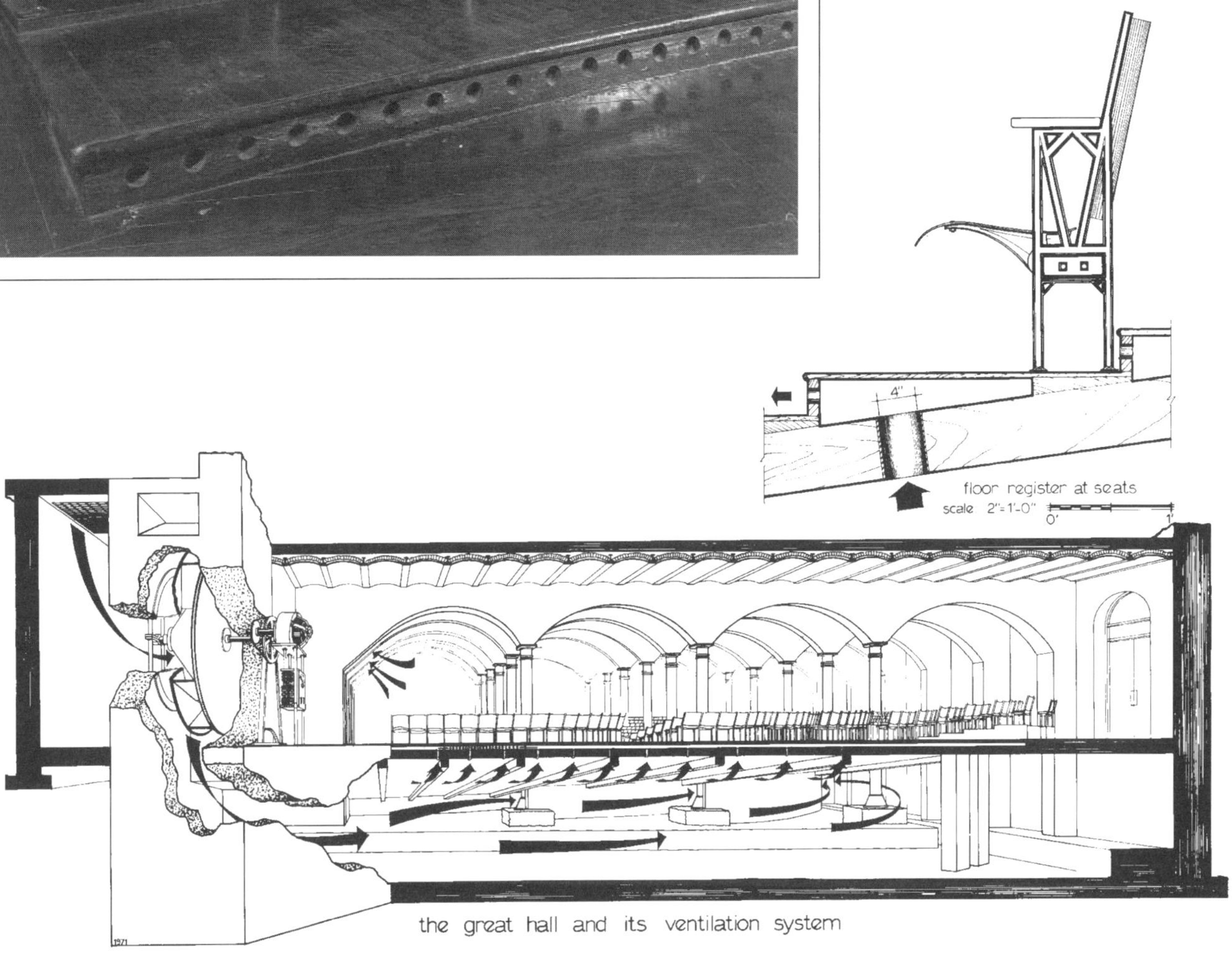

[FACING PAGE]

Harlem Fire Watch Tower, 1856, New York City, New York.

The Harlem Fire Watch Tower was constructed in 1856 by Julius B. Kroehl, following earlier examples fabricated by American iron pioneer James Bogardus. Bogardus may have been influenced by earlier iron-framed lighthouses.[4]

[THIS PAGE] *Martinsburg West Roundhouse, 1866, Baltimore and Ohio Railroad, Martinsburg, West Virginia.*

[RIGHT] *Interior view of cast-iron columns.*

[BELOW] *Half section of the roundhouse.*

The West Roundhouse is one of two identical buildings constructed by the Baltimore and Ohio Railroad in its Martinsburg, West Virginia, shop complex after the Civil War for locomotive repairs (to replace previous shop buildings destroyed during the war). This 16-sided structure is significant because of the cast-iron framing of the interior, which contains 16 work bays to house locomotives and machinery radiating from a central locomotive turntable. The high bell-shaped central part of the roundhouse was designed to act as a chimney to remove locomotive smoke.

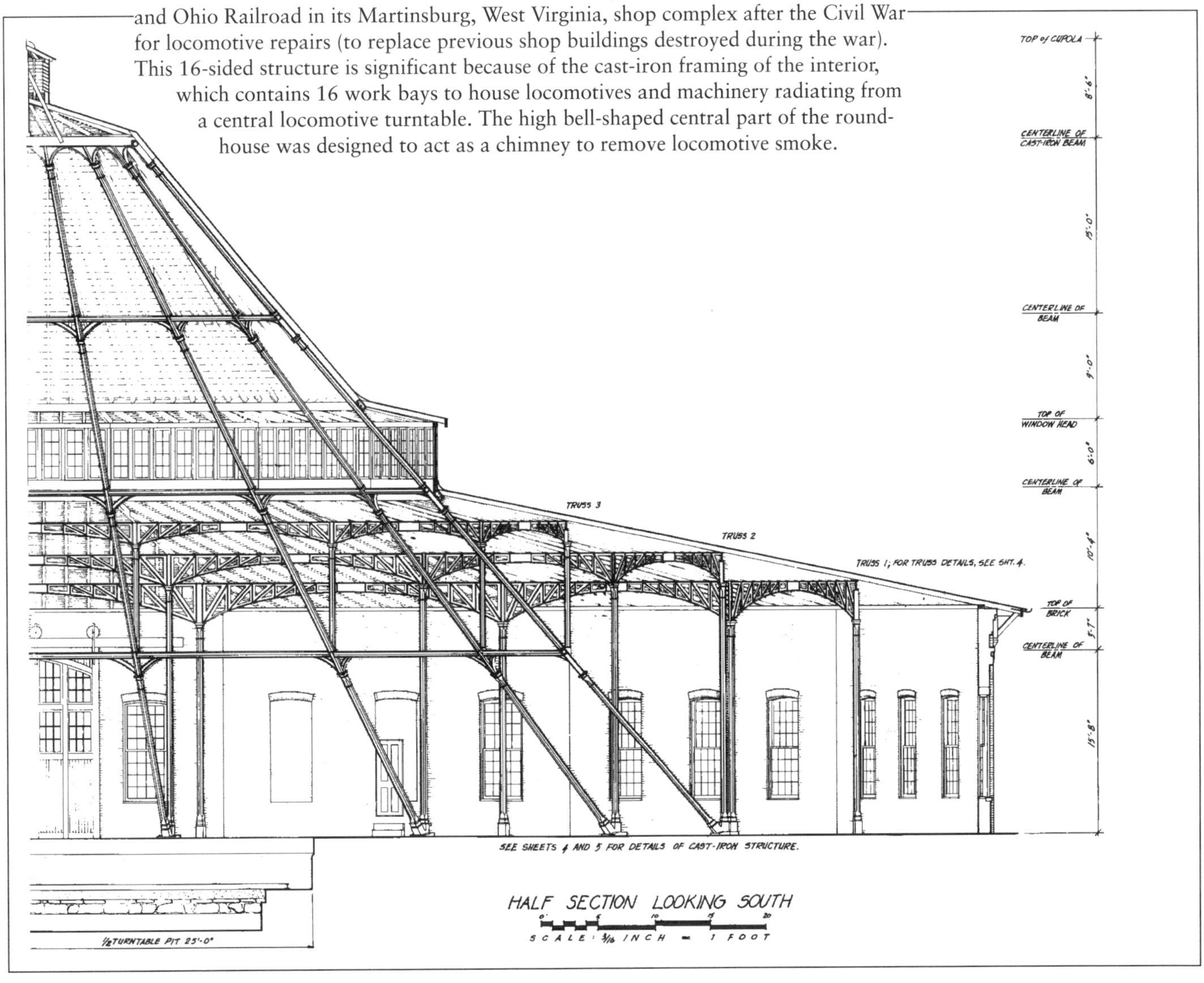

Mount Clare Passenger Car Shop, 1883-84, Baltimore and Ohio Railroad, Baltimore, Maryland.

The 22-sided Mount Clare Passenger Car Shop is an enormous industrial structure, 235 feet from side to side. Built in 1883-84 by the Baltimore and Ohio Railroad to build and repair passenger cars, the shop contains a 60-foot diameter turntable surrounded by 22 wrought-iron columns that support the roof. This structure is now the home of the Baltimore and Ohio Railroad Museum.

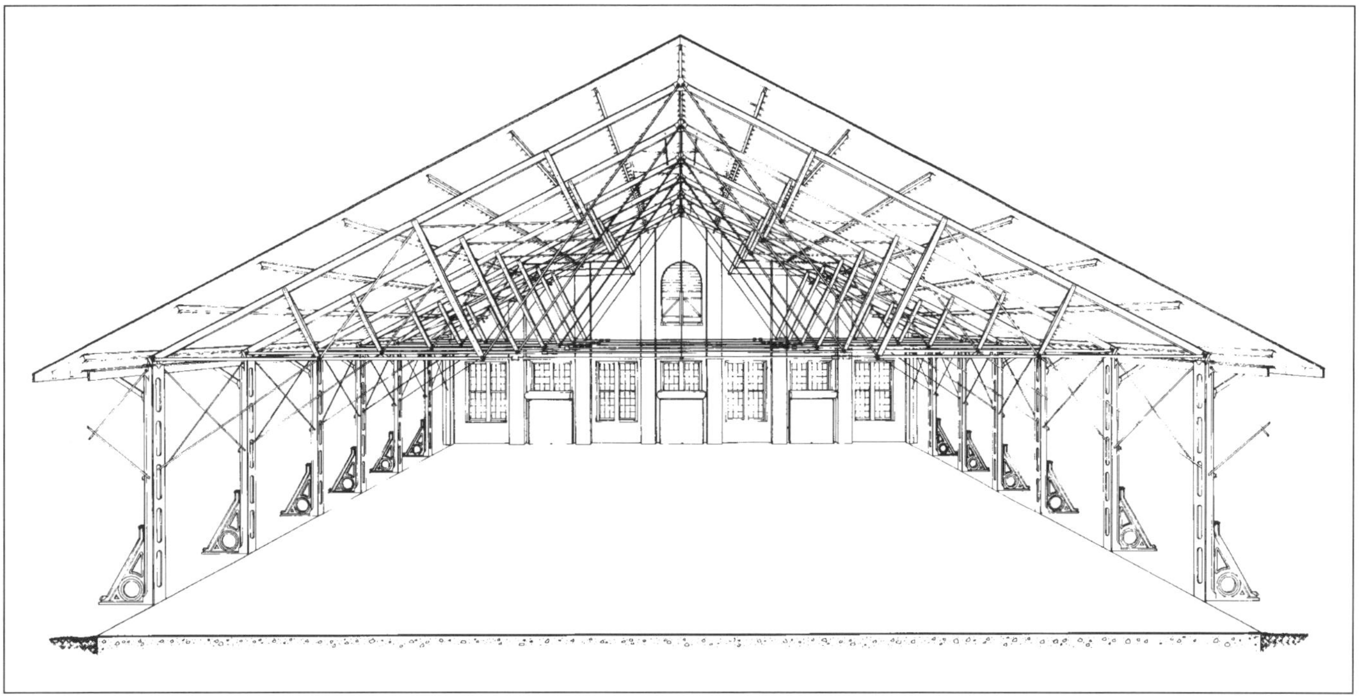

Bolt and Forge Shop, 1873, Baltimore and Ohio Railroad, Cumberland, Maryland.

[ABOVE] *Section of the shop.*

[RIGHT] *Exterior cast-iron brackets that buttress the columns.*

The Bolt and Forge Shop of the Baltimore and Ohio Railroad, built in Cumberland, Maryland, in 1873, was part of the Baltimore and Ohio Railroad's rolling mill complex in that city. These shops manufactured rails, bars, and other iron products. The cast-iron frame of the Bolt and Forge Shop is buttressed by unusual cast-iron brackets. The columns support a Fink roof truss of cast and wrought iron. All of the Baltimore and Ohio Railroad shops in Cumberland were demolished in 1981.

[ABOVE LEFT]
Composite roof truss of wood, cast iron, and wrought iron, Wheeling Freight Depot, 1852, Baltimore and Ohio Railroad, Wheeling, West Virginia.

[ABOVE RIGHT]
Trainshed showing steel columns and trusses composed of wood, and cast and wrought iron, 1883-85, Dearborn Station, Chicago and Western Railroad, Chicago, Illinois.

The third quarter of the nineteenth century was an era of great experimentation in structural materials, as wood and various forms of metal were used in a dizzying variety of combinations. The trusses of the Grafton Machine Shop and Foundry, for example, were composed of wooden chords, wrought-iron tie bars, and cast-iron struts.

Trainshed, Central of Georgia Railroad, 1861, Savannah, Georgia.

[LEFT] *Detail of truss connector in roof truss.*

[BOTTOM] *Gable roof.*

The Savannah trainshed of the Central of Georgia Railroad was built in 1861 as part of a comprehensive railroad terminal and shop complex. The roof trusses are tri-composite, composed of a wooden top chord, a wrought-iron lower chord, and cast-iron cruciform struts. An unusual feature of the truss is the radial connector joining the cast-iron struts with the lower chord. The trainshed was designed by Augustus Schwaab, a German immigrant who became an engineer for the Central of Georgia.

Montgomery County Jail, 1882, Crawfordsville, Indiana.

[RIGHT] *A pie-shaped cell.*

[BELOW] *Steel rotary cylinder in cellar, which rotated cells above.*

An unusual example of jail construction, this jail in Crawfordsville, Indiana, was composed of cells located in a circular two-level iron cage that rotated on a central pivot. There was only one opening for the cage, and the jailer used a hand-crank to rotate the structure to the cell he wanted. Supposedly, one prisoner kept foiling the rotation by sticking his wooden leg between the cage and his cell. The jail was an extreme fire hazard, but not until 1938 was the structure immobilized and openings made for each cell. The Montgomery County Jail was one of six similar rotary jails built in the Midwest in the 1880s.[5]

[LEFT] *Peavy-Haglin Grain Elevator, 1899-1900, St. Louis Park, Minnesota.*

The Peavy-Haglin Grain Elevator, constructed by Charles F. Haglin in 1899-1900, was the first circular reinforced concrete grain elevator in America, and the precursor of the form that the architect Le Corbusier called "the magnificent First Fruits of the new age."[6]

[ABOVE] *Alvord Lake Bridge, 1889, San Francisco, California.*

The Alvord Lake Bridge was the first bar-reinforced concrete bridge in America. Built by Ernest L. Ransome, a pioneer in the early use of concrete, the bridge contains twisted steel bars as reinforcement. Interestingly, the nature of the new building material was hidden somewhat, as the surface of the concrete was textured to resemble blocks of sandstone, possibly to match sandstone structures in a nearby children's playground. The soffit was further decorated with concrete resembling stalactites.

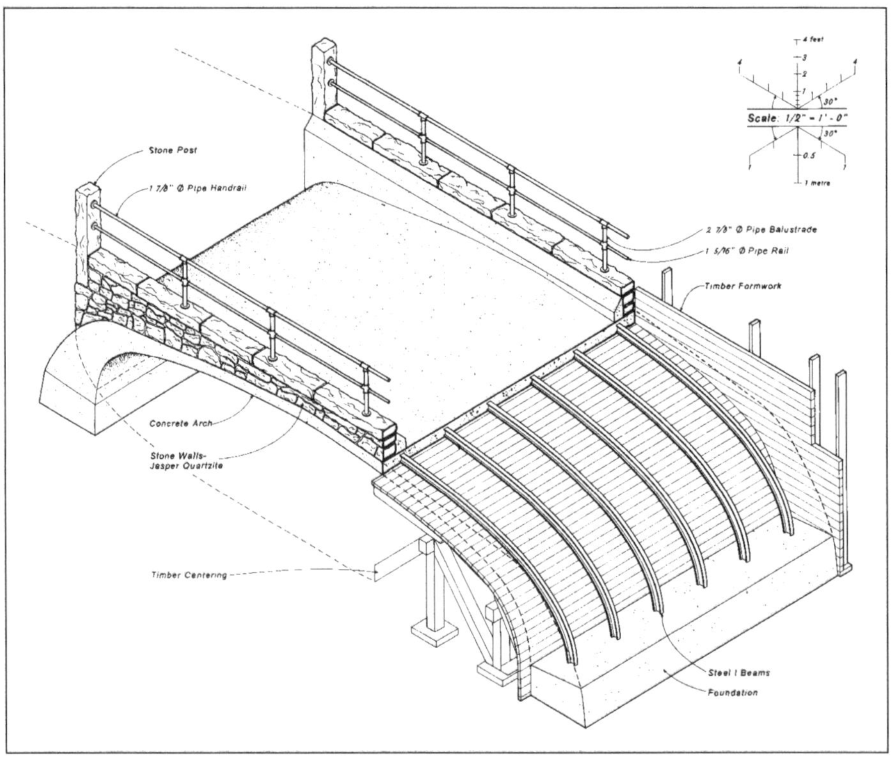

Melan Arch Bridge, 1894, Rock Rapids, Iowa.

[ABOVE] *Elevation of bridge.*

[RIGHT] *Isometric drawing showing the reinforced concrete system used in construction.*

The Melan Arch Bridge in Iowa was the first use in America of the reinforced concrete system designed by Josef Melan of Austria, and was built by Melan's representative in America, Fritz von Emperger. The bridge's concrete was reinforced by 4-inch steel I-beams, bent to the shape of the arch. In 1964, the bridge was moved 4 miles to save it from destruction.

A Different Path: Vernacular Technology

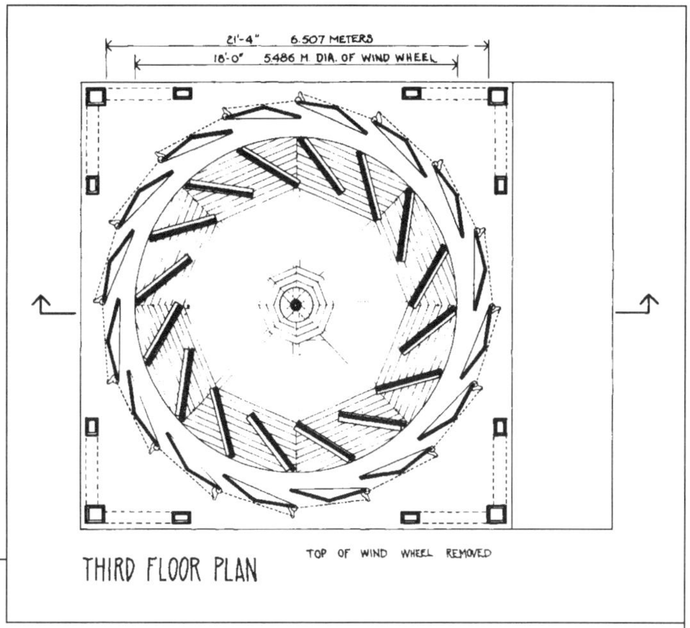

Gladden Wind Turbine, 1890-91, East Randolph, New York.

George Gladden, a New York farmer who had observed a vertical wind turbine on a farm in Nebraska, obtained the rights to construct a similar one on his farm in 1890. The wind wheel, 13-feet high and 18 feet in diameter, sits on the top story of a four-story wood-frame mill. The wheel contains 16 shutters and was capable of producing as much as 20 horsepower. The turbine originally powered machinery for elevating and grinding grain, an apple grater and press for making cider, a wood lathe, and a shop for general farm repair work.

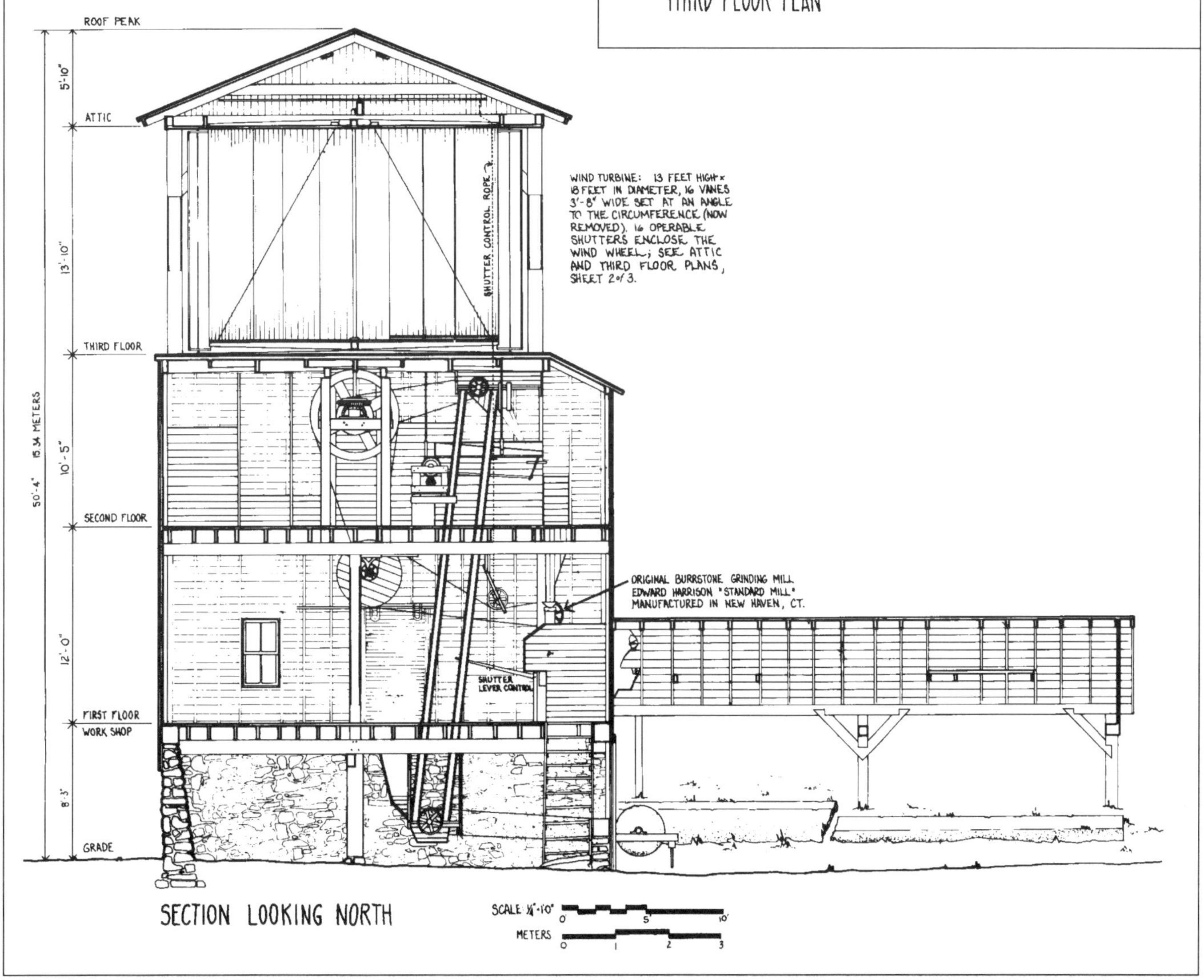

[TOP RIGHT]
Tyringham Shaker Settlement Dam, 1830, Tyringham, Massachusetts.

[BOTTOM RIGHT]
Laundry equipment, Shaker North Family Laundry and Woodstore Building, Mount Lebanon, New York.

Technology played a central role in the community life of the Shakers, but not in the same way as elsewhere in the country. Shakers invented labor-saving devices and used mechanization mostly to improve handcraft technologies, rather than as a means to increasingly replace hand labor. The Tyringham Shaker Settlement in Massachusetts contained a sawmill, a gristmill, a furnace, a blacksmith shop, and a planing mill. The dam, constructed in 1830, created a pond to provide water power for these industries.[7] The laundry equipment of the Mount Lebanon North Family included a Shaker-built centrifugal dryer (behind the stove) and a washing machine (on the right), both powered by a "Backus 1872" Turbine Water Motor.

Avery-Bartholomew Patent Railroad Iron Bridge, c. 1877, Groton, New York.

The third quarter of the nineteenth century was a time of great experimentation in American bridge fabrication. One of the more unusual designs was this bridge in Groton, New York, based on an 1877 patent granted to Oliver Avery, Jr., and Caleb Bartholomew, which specified railroad rails for the upper and lower chords. This was a short-lived design, but it helped start the Groton Iron Bridge Company, which became the leading fabricator of metal truss bridges in New York.

Coleman Bridge, c. 1894, Windsor, Massachusetts.

In western New England in the 1890s, Charles Henry Ball, a mechanic and inventor from East Windsor, Massachusetts, built at least 25 bridges based on his patent for iron pipe bridges. Ball's goal, he later wrote, was "to produce a strong, cheap bridge, that would last as long as any iron bridge, and cost but little, if any, more than a good wooden bridge."[8] Coleman Bridge in Windsor, Massachusetts, is the only one of Ball's bridges still standing.

Unidentified suspension bridge, c. late nineteenth century, Texas.

Several vernacular suspension bridges appeared in Texas in the late nineteenth century, using iron pipes, cable stays, and other innovative features.

E. and T. Fairbanks and Company, c. 1876, St. Johnsbury, Vermont.

[RIGHT] *Two-level, two-aisle covered bridge.*

[BELOW] *Town lattice truss and reinforcing arch of the covered bridge.*

E. and T. Fairbanks and Company, makers of industrial scales, constructed a two-story covered bridge around 1876 to connect buildings on opposite sides of the Sleepers River in St. Johnsbury, Vermont. This type of truss was first patented by Ithiel Town in 1820. The heavy laminated arches may be later additions. The bridge and other company structures were destroyed by fire in 1972.

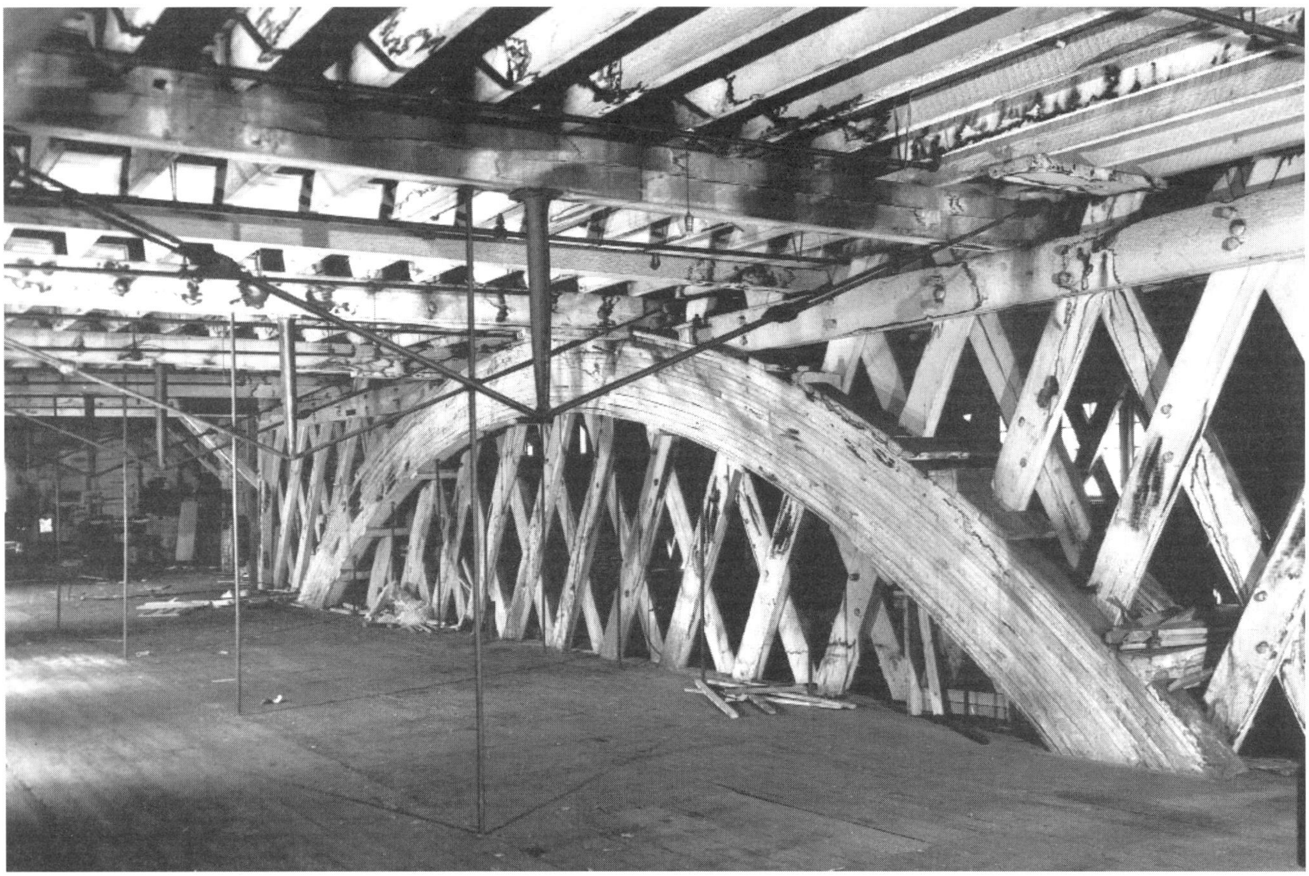

Goat Trail, 1895, near Newhalem, Washington.

[ABOVE] *The formidable landscape through which Goat Trail was constructed.*

[LEFT] *Remains of Bridge No. 2 and overhangs where rock was blasted out.*

The Goat Trail is a testament to the power of gold. Washington's Upper Skagit region was a harsh environment in the nineteenth century, "consisting of deep, impassible gorges, lofty divides, and snow-capped peaks..."[9] But gold was discovered there in 1879, and where there's a will, there's a way. An extremely hazardous trail, called the Goat Trail, was constructed to the mining region, but huge granite rock outcrops at a place called Devil's Corner hindered the journey. Local residents, finally tired of waiting for the state to take action, raised money themselves and blasted half-tunnels through the outcrops in 1895. A railroad constructed along the Skagit River in the 1920s finally eliminated the need for the Goat Trail.

Hurricane Irrigation Canal, 1894-1904, near Hurricane, Utah.

[ABOVE LEFT] *Portions of the canal were carved through rock cliffs.*

[ABOVE RIGHT] *Water control gates at head of canal.*

To provide water for new agricultural land, settlers in southern Utah began construction of the Hurricane Irrigation Canal in 1894. Using only shovels, picks, and hand-drills, these men worked on the canal for 10 years, alternately farming in the summer months and digging the canal in the winter months. The 7-mile canal was constructed through precipitous canyon land, requiring nine tunnels. It finally was finished in 1904, providing water for 2,000 acres of land.[10]

A CLOSER LOOK

Engineering Embellished: Design & Ornamentation

Although form usually followed function in nineteenth-century American engineering and technology, decoration and ornamentation often accompanied function. Whether intended to alleviate fears about machinery and technology, assert the central role of engineering and technology in American life, or apply to technology the same aesthetic treatment accorded other areas of contemporary life, notes the historian John Kasson, "form followed... fashion and symbolic expression to a surprising degree in nineteenth-century American industrial design..."[11]

Decorated frame of spinning jack, c. 1870, Watkins Woolen Mill, Lawson, Missouri.

A CLOSER LOOK

Engineering Embellished: Design & Ornamentation

[ABOVE] *Gothic arch, Poinsett Bridge, 1820, near Tigerville, South Carolina.*

Although the Poinsett Bridge was built in a remote part of the South Carolina upcountry in 1820 by the state's Board of Public Works, the bridge was given a decorative arch. The Poinsett Bridge was part of a state road that ran between Charleston and Greenville, South Carolina, and on to Asheville, North Carolina.

[ABOVE] *Engine house, McCulloch Gold Mill, 1832, near Jamestown, North Carolina.*

Local tradition states that slaves erected this building under the supervision of Elizier Kersey, a Cornish emigrant. The Gothic arch was probably more than simply decorative, but its precise function is now a mystery.

A CLOSER LOOK

Engineering Embellished: Design & Ornamentation

[LEFT TOP] *Top of standpipe tower, Louisville Water Company's Pumping Station No. 1, 1856 (partially rebuilt, 1890), Louisville, Kentucky.*

[LEFT BOTTOM] *Detail of a column, Petersburg Gas Light Company Gas Holder, 1876, Petersburg, Virginia.*

[ABOVE] *Tower, Clinton Mill, 1849 (tower roof, 1853), Woonsocket, Rhode Island.*

A CLOSER LOOK

Engineering Embellished: Design & Ornamentation

[ABOVE] *Boston Public Garden Suspension Bridge, 1867, Boston, Massachusetts.*

An early collaborative effort between architect William Preston and engineer Clemens Hershell, this bridge is an elegant example of public engineering.[12]

[RIGHT] *Gothic Arch Bridge (Central Park Bridge No. 28), 1864, New York City, New York.*

References

1. Carl W. Condit, *American Building Art: The Nineteenth Century* (New York: Oxford University Press, 1960) p. 3.

2. Carl Condit, "Buildings and Construction," in Melvin Kranzberg and Carroll W. Pursell Jr., eds., *Technology in Western Civilization, Vol. I* , (New York: Oxford University Press, 1967) pp. 368–373.

3. Condit, "Buildings and Construction," pp. 372–373.

4. Condit, *American Building Art: The Nineteenth Century*, p. 282, n.7.

5. Walter A. Lunden, "The Rotary Jail, or Human Squirrel Cage," *Journal of the Society of Architectural Historians*, XVIII (1959):149–157; Connie Riggs, "The Human Squirrel Cage," *Indian History Bulletin,* 53 (1975):123–126.

6. Daniel L. Schodek, *Landmarks in American Civil Engineering* (Cambridge, MA: MIT Press, 1987) p. 283.

7. Brooke Hindle and Steven Lubar, *Engines of Change: The American Industrial Revolution, 1790–1860* (Washington, D.C.: Smithsonian Institution Press, 1986) p. 239.

8. Lola Bennett, "Coleman Bridge," U.S. Department of the Interior, Historic American Engineering Record (HAER), No. MA-119, Prints and Photographs Division, Library of Congress, Washington, D.C., (1990) p. 5.

9. Kenneth D. Rose, "Goat Trail [Mining Road] at Devil's Corner," U.S. Department of the Interior, Historic American Engineering Record (HAER), No. WA-19, Prints and Photographs Division, Library of Congress, Washington, D.C., (1987) p. 2.

10. Andrew Karl Larson, *I Was Called to Dixie* (Salt Lake City, UT: Deseret News Press, 1961) pp. 382–402.

11. John F. Kasson, *Civilizing the Machine: Technology and Republican Values in America, 1776–1900* (New York: Penguin, 1976) p. 154.

12. Peter Stott, *A Guide to the Industrial Archeology of Boston Proper* (Cambridge, MA: MIT Press, 1984) pp. 17–18.

4

Going Places

"If God had designed that His intelligent creatures should travel at the frightful speed of 15 miles an hour by steam, He would have foretold it through His holy prophets. It is a device of Satan to lead immortal souls down to Hell."[1] So declared a local school board in Ohio in 1828 in reaction to the idea of a railroad. Although Americans generally embraced technology in the nineteenth century, there was also ambivalence when any particular technology first appeared. As a growing national economy began to encourage Americans to look beyond their local borders, people discovered the inadequacies of the country's fledgling transportation network. Before the vast transportation improvements in the early decades of the nineteenth century, in fact, the shortest distance between two points in America was, alas, usually not a straight line. Farmers in western Massachusetts found this out, to take a case in point. Even though the town of Northampton was located only a hundred miles west of Boston, it was cheaper for the region's farmers to send grain to their capital city on an amazingly circuitous route, first overland 36 miles south to the Connecticut River, then another 250 miles down that river, through Long Island Sound, around Cape Cod, and finally on to Boston. The brief land portion of this journey, moreover, was 13 times more expensive than the water portion.[2]

This sort of experience encouraged the building of turnpikes, canals, and railroads, Satanic or otherwise. New transportation networks were constructed in the early years of the nineteenth century as the country sought to unite its disparate geographic elements and to create new markets for its young manufacturing concerns. Roads and turnpikes were carved through the wilderness, a canal craze swept the nation after the completion of the Erie Canal in 1825, steamboats plied the western waters, and a railroad mania after the 1830s resulted in America having by 1900 almost half of all the railroad mileage in the world.[3]

Railroads in particular were crucial in knitting the country together. So pervasive was the iron horse that a French visitor to America commented as early as the 1830s on the young country's penchant for building railroads:

> The Americans have railroads in the water, in the bowels of the earth, and in the air… When they cannot construct a real, profitable railway from river to river, from city to city, or from State to State, they get one up, at least as a plaything or until they can accomplish something better….[4]

By 1850, railroad mileage in the United States was almost triple that of canals, and the railroad quickly eclipsed the canal as a major freight hauler. In 1852, for example, freight could be shipped from Cincinnati,

[FACING PAGE]

Omaha Bridge, 1885-87 (photographed, 1905), Omaha, Nebraska.

Omaha Bridge was one of the many long-span railroad bridges built over the great Midwestern rivers by George S. Morison in the late nineteenth century, facilitating the westward expansion of the railroad.

MISSOURI
RIVER

[ABOVE] *New Portland Suspension Bridge, c. 1866, New Portland, Maine.*

Ohio, to New York City by canal in 18 days, but in only 6 to 8 days by railroad.[5]

The "transportation revolution" of the nineteenth century also resulted in the building of hundreds of bridges, aqueducts, and tunnels, as the country's numerous rivers and mountain chains were conquered. Endless experimentation and variation in technology, form, design, and ornamentation marked the construction of these structures, and bridges in particular came to symbolize the art, power, and skill of engineering. As early as 1823, a visitor to the United States remarked that Americans "build much better bridges than Parthenons and Capitols."[6]

By Water: Canals and Maritime

[LEFT] *Inclined Plane No. 9 West, Morris Canal, 1825-31 (photograph, undated), near Port Warren, New Jersey.*

A partially submerged cradle, which carried canal boats up and down the plane, is in foreground.

The Morris Canal was constructed between 1825 and 1831 across northern New Jersey to connect the Hudson River with the Delaware River. Much of the terrain traversed was hilly and rugged, requiring the use of 23 inclined planes. To ascend a plane, a canal boat floated into a cradle, which then was pulled up the plane on a track by a chain, which was powered by a water turbine. Inclined Plane No. 9 West was the largest of the canal's planes, with a rise of 100 feet and a length of more than 1,500 feet.[7]

[RIGHT] *Waterford Locks, 1822-26, Champlain Canal, Waterford, New York.*

In 1817, as work began on the Erie Canal, work also was progressing on the Champlain Canal, designed to connect Lake Champlain with the Erie Canal and the Hudson River. The canal was completed to Waterford, New York, by 1822, where it connected with the Hudson River through these three locks.

[TOP RIGHT]

Paw-Paw Tunnel, 1836-48, Chesapeake and Ohio Canal, near Old Town, Maryland.

[BOTTOM RIGHT]

Monocacy Aqueduct, 1829-33, Chesapeake and Ohio Canal, near Dickerson, Maryland.

The Chesapeake and Ohio Canal was planned to follow the Potomac River as a route through the Appalachian Mountains. Construction on the "Great National Project" began on July 4, 1828. Sections opened as they were completed, but the canal never reached the Ohio River, stopping in Cumberland, Maryland, in 1850. Engineering features of the canal included the brick-lined Paw-Paw Tunnel, 3,118-feet long, and the Monocacy Aqueduct, a seven-arch structure 433-feet long and the longest of the canal's 11 aqueducts.

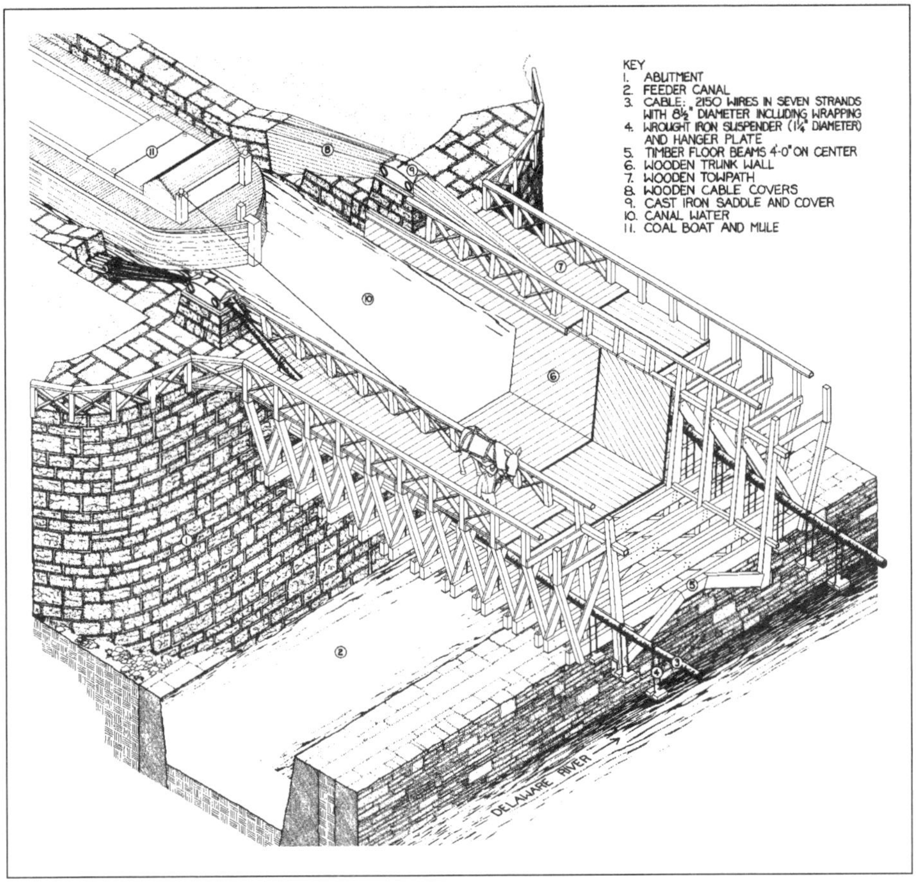

Delaware Aqueduct, 1847-1849, Lackawaxen, Pennsylvania, and Minisink Ford, New York.

[LEFT] *Isometric of the aqueduct.*

[BELOW] *The aqueduct after its conversion to a road bridge.*

The Delaware Aqueduct, built in 1847-49, is the oldest suspension bridge standing in America. It was designed by John Roebling as part of the improvement of the Delaware and Hudson Canal, originally constructed between 1825 and 1829. Each of the two cables is composed of 2,150 iron wires, bound in seven strands. Roebling later incorporated many of the same methods he used for the Delaware Aqueduct in the plans for the Brooklyn Bridge.

[ABOVE] *Sand Key Lighthouse, 1853, Sand Key, Florida.*

[LEFT] *A cast-iron footing.*

The Sand Key Lighthouse is an iron screw-pile lighthouse, with submerged wrought-iron piles screwed into sand and coral. The 132-foot-high tower was designed in 1853 by noted lighthouse designer I.W.P. Lewis, and built under the supervision of Lt. George C. Meade of the Bureau of Topographical Engineers, later a Union general during the Civil War.

[FACING PAGE]

Seven-Foot Knoll Lighthouse, 1856, near Riviera Beach, Maryland.

The Seven-Foot Knoll Lighthouse, built in 1856 by the United States Lighthouse Board, is made up of riveted wrought-iron plates mounted on cast-iron screw-piles. The screw-piles are original, but the wrought-iron "house" replaced the original cast-iron square house at some time in the late nineteenth century. This lighthouse controlled shipping in the Chesapeake Bay, and is an early example of a screw-pile-type structure, which used cast-iron piles to avoid heavy masonry foundations.

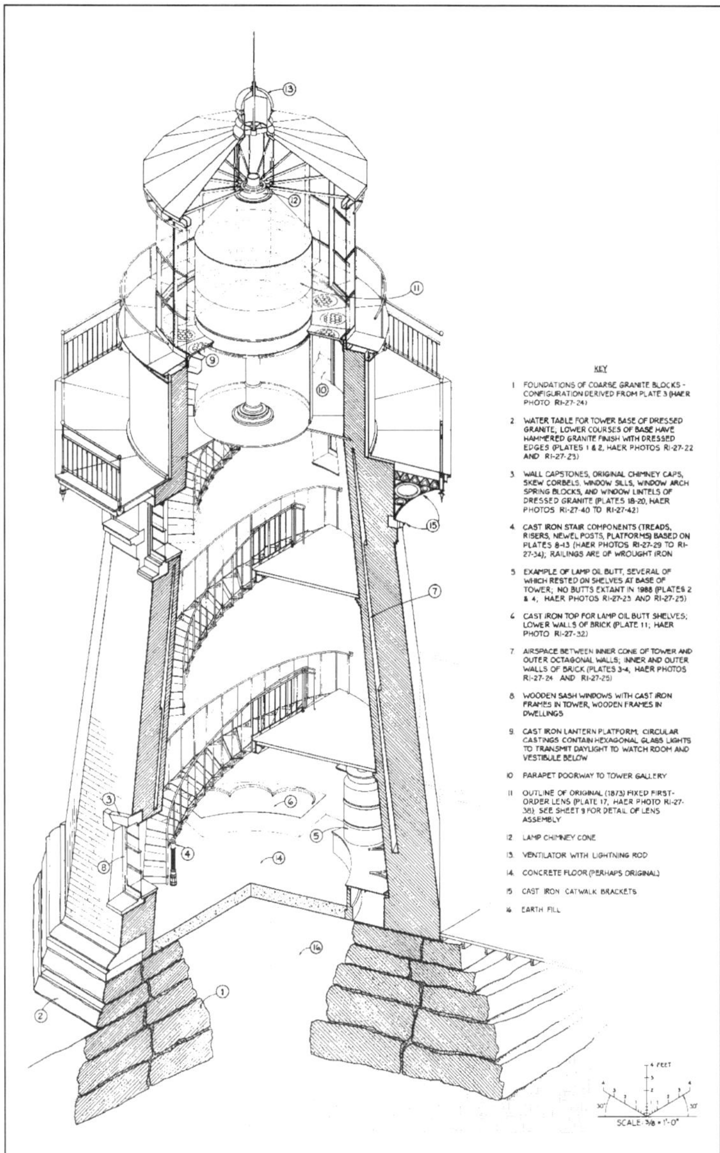

[LEFT] *Block Island Southeast Light, 1875, Block Island, Rhode Island.*

[BELOW] *Lighthouse and light keeper's quarters.*

Block Island is the first landfall for ships approaching the New England coast from the south or southeast, and the United States Lighthouse Board constructed one of the best-equipped lighthouses on the east coast for this significant site in 1875. The Block Island Southeast Light initially was equipped with a top-of-the-line first-order Fresnel lens and a steam siren fog signal. Joseph Henry, Chairman of the Lighthouse Board and Secretary of the Smithsonian Institution, conducted experiments on fog-signal sound phenomena using the Block Island fog signal station soon after it began operation.

Schooner Wawona, *1897, Seattle, Washington.*

The *Wawona* is a representative example of the many American wooden schooners built in the late nineteenth century to transport lumber from the Pacific Northwest to other parts of the west coast. The three-masted *Wawona* was constructed in 1897 in the Fairhaven, California, yards of Hans D. Bendixsen, a prominent local shipbuilder.

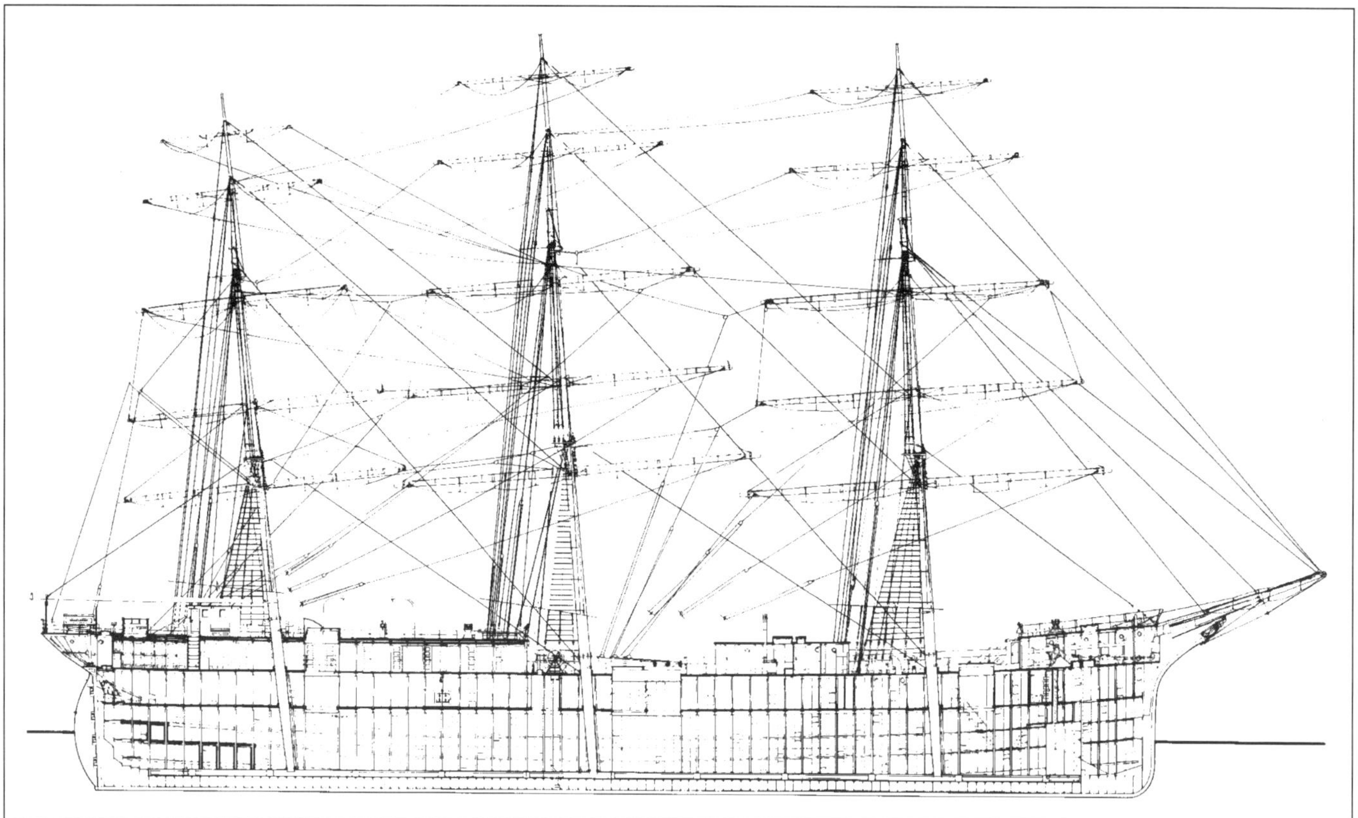

Balclutha, *1886,*
San Francisco, California.

[ABOVE] *Inboard profile.*

[RIGHT] *Steering gear cover (to the right), saloon skylight, and chart house (in background).*

The *Balclutha* was built in 1886 in Glasgow, Scotland, and is one of the last surviving full-rigged steel-hulled ships of the nineteenth century. The *Balclutha* was used in the merchant trade until it was sold in 1899, after which it plied the Pacific coast of the United States hauling grain, lumber, and salmon. The *Balclutha* made her last voyage in 1930 and is now berthed in the San Francisco Maritime National Historical Park.

The Iron Horse: Railroads

"Three-Spot" Locomotive, c. 1870s, Duluth and Iron Range Railroad, Two Harbors, Minnesota.

This Mogul-type locomotive was built by the Baldwin Locomotive Works of Philadelphia, Pennsylvania, probably in the 1870s, and was shipped by barge in 1883 to what would become Two Harbors, Minnesota. The northern Minnesota iron range then was being developed, and the "Three-Spot," so named because it was Engine No. 3, helped build the first railroad to the iron mines by hauling rails and supplies along the line.

Granite Railway Incline, 1826, Quincy, Massachusetts.

[RIGHT] *Track and pulleys of the original quarry incline.*

[BELOW] *Reconstructed railcar.*

The Granite Railway, one of the first railways in America, was constructed in 1826 to transport stone for the building of the Battle of Bunker Hill monument in Boston, Massachusetts. An incline 315-feet long was built to gain access to Pine Hill Quarry. Cars on the incline carrying granite blocks were raised and lowered by an endless chain.[8]

Trainshed, Union Station, 1897-98, Montgomery, Alabama.

The Union Station Trainshed, constructed in Montgomery, Alabama, by the Louisville and Nashville Railroad between 1897 and 1898, is a throwback to an earlier era. By the late 1890s, the gable-roofed trainshed had given way to the larger balloon shed elsewhere, but because of the small number of tracks coming into Montgomery, the Louisville and Nashville Railroad determined that the older form would suffice. The Pratt trusses of the shed are also of an older type, composite trusses of timber, wrought iron, and cast iron. Most structures used all-metal trusses by this date. Another unusual feature of these trusses is the use of a Phoenix column as the center strut in the top chord, indicating that salvaged bridge members may have been used in the construction of the trusses.[9]

[THIS PAGE]

[ABOVE] *Vestal Passenger and Freight Station, 1881, Delaware, Lackawanna, and Western Railroad, Vestal, New York.*

[LEFT] *Middletown Station, 1896, Erie Railroad, Middletown, New York.*

[FACING PAGE]

Point of Rocks Station, 1875, Baltimore and Ohio Railroad, Point of Rocks, Maryland.

As part of the dominant transportation mode of the nineteenth century, with both stylistic and technological requirements, railroad stations became experiments in attempts to unite architecture and engineering. John Ruskin, calling for a primarily functional approach to station design, argued at mid-century that, "Railroad architecture has or would have a dignity of its own if it were only left to its work." Many stations were indeed built without regard to ornament or style, but as Carroll Meeks has shown, even small station design by the 1870s emphasized the attributes of the sublime and the picturesque in a style Meeks called "picturesque eclectic."[10] These three stations demonstrate a variety of influences, from Gothic and Queen Anne to Orientalism and Romanesque.

POINT OF ROCKS

Savannah Repair Shops, Central of Georgia Railroad, Savannah, Georgia.

[RIGHT] *Combination smokestack, water tank, and privy.*

[BELOW] *Base of smokestack, showing brick privies and cast-iron water tank.*

The Central of Georgia Railroad shop site in Savannah, Georgia, contains some of the oldest surviving buildings of any railroad shop site in the United States and includes this unique combination smokestack, water tank, and privy. The 125-foot-tall tower was constructed in 1855 to exhaust smoke from several shops and to store water for locomotives and a stationary steam engine boiler. Buttresses at the bottom of the stack formed cells that were used by workmen as privies. The 16-sided cast-iron water tank, which encircles the stack, was manufactured by D. & W. Rose of Savannah.

Interlocking System, South Station, 1899, Boston, Massachusetts.

[ABOVE] *Semaphore signals on Signal Bridge No. 7.*

[RIGHT] *Interlocking machine.*

After its opening in 1899, South Station in Boston, Massachusetts, became one of the nation's busiest railroad stations, handling as many as 700 trains per day. A Westinghouse electro-pneumatic interlocking system, consisting of semaphore signals, signal bridges, and switches controlled from three interlocking towers, was installed to facilitate this heavy traffic. The system allowed operators to control the complex movement of trains through switches and signals that were interconnected and operable only in sequence, greatly reducing the possibility of human error.

[BELOW] *Part of the track, Third Avenue Elevated Railway, begun in the 1880s, New York City, New York.*

[RIGHT] *Fordham Road Station, Third Avenue Elevated Railway.*

As population pressures pushed New York City's boundaries northward, first horse cars and then elevated railroad systems were built to connect the expanding metropolis. The elevated railroad system gave millions of New Yorkers the option of mobility, but trains were slow, and everyone disliked the intrusive and sun-blocking elevated structure. Work began on New York's next rapid-transit experiment, the subway, in 1900.

Union Elevated Railroad (Union Loop), 1897, Chicago, Illinois.

Chicago, Illinois, followed New York's example of elevated railroad systems and sought to rationalize the jumble of competing elevated lines by creating in the late 1890s a "union" loop in the downtown section of the city. The Union Elevated Railroad joined the systems of four separate elevated railroad companies and gave downtown Chicago its nickname, the Loop.

Blue Ridge Tunnel, 1850-1858, near Afton, Virginia.

At 4,273 feet, the Blue Ridge Tunnel was the longest railroad tunnel in the world when completed in 1858 for the Blue Ridge Railway. The tunnel was constructed by Claudius Crozet (1790-1864), a brilliant French engineer who emigrated to the United States in 1816 and who had taught at the United States Military Academy at West Point and served as the Principal Engineer of the Virginia Board of Public Works. In 1850, Crozet started two crews boring on either side of the Blue Ridge Mountains, using only contemporary hand-drills, pickaxes, and black powder. Although the crews were almost a mile apart and had to negotiate a curved alignment, the two sides were only half an inch off line when they met on Christmas Day in 1856.[11]

A CLOSER LOOK

East Broad Top Railroad

The East Broad Top Railroad, constructed in the early 1870s as a narrow-gauge coal carrier in Huntingdon County, Pennsylvania, remains remarkably intact as the heart of a historic transportation and industrial system that includes the still-extant right-of-way, rolling stock, repair shops, bridges and tunnels, stations, workers' communities, and remnants of the supporting metal and mining industries. This National Historic Landmark's shop complex, located in Rockhill Furnace, made the East Broad Top Railroad virtually self-sufficient in servicing its line, rolling stock, and facilities.

[BELOW] *Steam locomotives in the roundhouse, East Broad Top Railroad shop complex, Rockhill Furnace, Pennsylvania.*

[ABOVE] *Triple stub switch, controlled by a hand-operated switch stand, which changed track alignment, East Broad Top Railroad.*

A CLOSER LOOK

East Broad Top Railroad

[ABOVE] *Water column, which could swing over the track to line up with the locomotive tender's water hatch, East Broad Top Railroad.*

[RIGHT TOP] *Timber-frame jib crane, 1878, in foundry of the East Broad Top Railroad.*

[RIGHT BOTTOM] *Interior, Wray's Hill Tunnel, East Broad Top Railroad.*

A CLOSER LOOK

East Broad Top Railroad

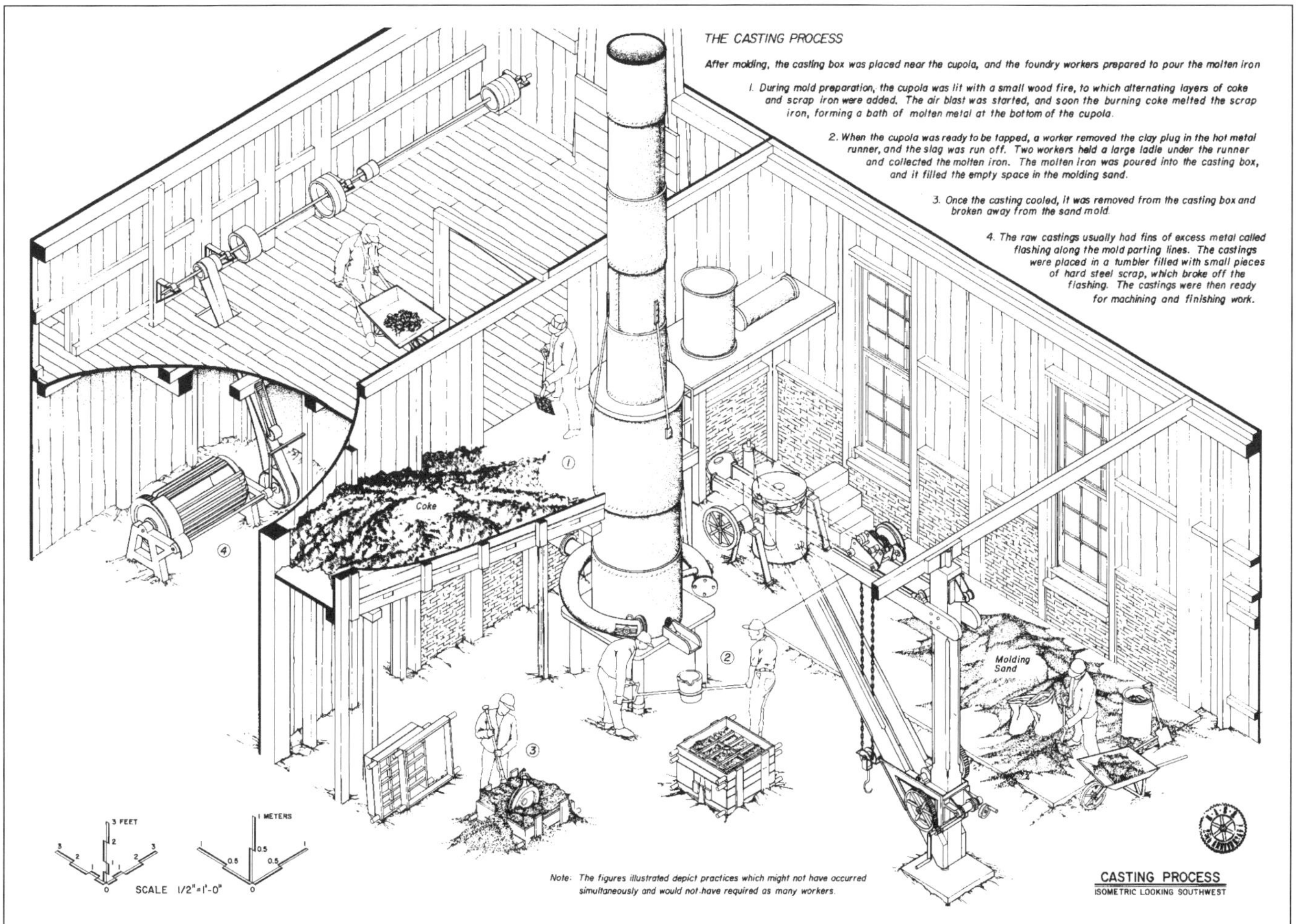

The process of making an iron casting, in foundry of the East Broad Top Railroad.

Spanning the Century: Bridges

[ABOVE] *Humpback Covered Bridge, 1835 (possibly 1857), Covington, Virginia.*

[FACING PAGE]

[TOP] *Wilson Bridge, 1819, near Hagerstown, Maryland.*

[BOTTOM] *Dunlaps Creek Bridge, 1836-1839, Brownsville, Pennsylvania.*

As Americans pushed westward throughout the nineteenth century, new bridges were built to cross the country's many rivers. These three bridges show the variety of materials with which bridges could be constructed. The Wilson Bridge was constructed in 1819 as part of a local turnpike intended to feed into the National Road, which began 60 miles westward in Cumberland, Maryland. Stone bridges were advocated originally by the Secretary of War in the plans for all roads leading to the National Road, and locals and the state insisted on stone when Army engineers later recommended less-expensive wooden bridges. The five-arch bridge was built by Silas Harry of Pennsylvania. Twenty years later, a new material appeared in a bridge on the National Road itself. Dunlaps Creek Bridge in Brownsville, Pennsylvania, was the country's first all-iron bridge. This arch was constructed between 1836 and 1839 by Richard Delafield of the United States Army Corps of Engineers. The Humpback Bridge, a covered wooden bridge, also was built as part of a transportation route to the West, for the James River and Kanawha Turnpike. Its multiple king-post trussing system with curved upper and lower chords is a rare form in America.

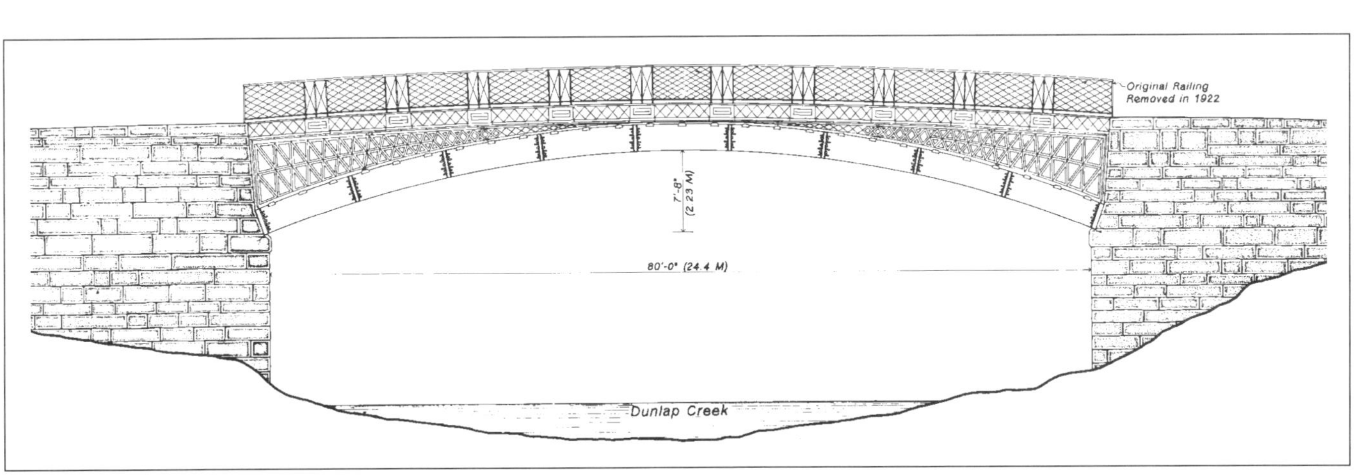
Original Railing
Removed in 1922
7'-8"
(2.23 M)
80'-0" (24.4 M)
Dunlap Creek

Montague Covered Bridge, 1870, Montague, Massachusetts.

The Montague Covered Bridge was 770-feet long, one of the longest covered bridges in the country until it was destroyed in a hurricane in 1936. Charles Dickens described a ride through a similarly long covered bridge in Pennsylvania: "It was profoundly dark; perplexed, with great beams crossing and recrossing it at every possible angle; and through the broad chinks and crevices of the floor the rapid river gleamed, far down below, like a legion of eyes."[12]

[ABOVE] *Carrollton Viaduct, Baltimore and Ohio Railroad, 1829, Baltimore, Maryland.*

The Carrollton Viaduct is the oldest surviving railroad bridge in the United States, constructed for the Baltimore and Ohio Railroad in 1829 by Jonathan Knight and Caspar Wever. William Gibbs McNeill and other U.S. Army engineers who had assisted with the initial survey for the Baltimore and Ohio Railroad line also had a hand in the construction of the viaduct.

[RIGHT] *Canton Viaduct, 1835, Canton, Massachusetts.*

William Gibbs McNeill followed his experience on the Carrollton Viaduct 6 years later by constructing the Canton Viaduct in Canton, Massachusetts. This massive granite structure, built for the Boston and Providence Railroad, is more than 600-feet long.

[THIS PAGE]

Reading-Halls Station Bridge, c. 1846, near Muncy, Pennsylvania.

[ABOVE] *Bridge panels showing the Egyptian Revival decorative motif.*

[RIGHT] *Cast-iron compression members.*

One of the first all-metal railroad bridges in America was erected around 1846 for the Philadelphia and Reading Railroad in eastern Pennsylvania. Composed of cast-iron diagonal compression members and wrought-iron top and bottom chords, the Reading-Halls Station Bridge is adorned with Egyptian Revival decorative motifs. The bridge was constructed by engineer Richard B. Osborne as a Howe truss, a popular form at the time for wooden bridges. Osborne designed an all-metal bridge in part because boatmen on the competing Schuylkill Canal kept burning down the railroad's timber bridges.[13] The bridge later was moved to its present location over a railroad and converted into a vehicular bridge.

[FACING PAGE]

Starrucca Viaduct, 1848, Lanesboro, Pennsylvania.

Built by the New York and Erie Railway in 1848 and more than 1,000 feet in length, the Starrucca Viaduct was one of the earliest major links between the eastern seaboard and the West. Structurally, the bridge was both a descendent of established form as well as a forerunner of future developments. Soon after the completion of the Starrucca Viaduct, stone construction was superseded by new developments in iron and steel, but this viaduct also contains footings that are massive blocks of concrete, one of the first uses of structural concrete in an American bridge.[14]

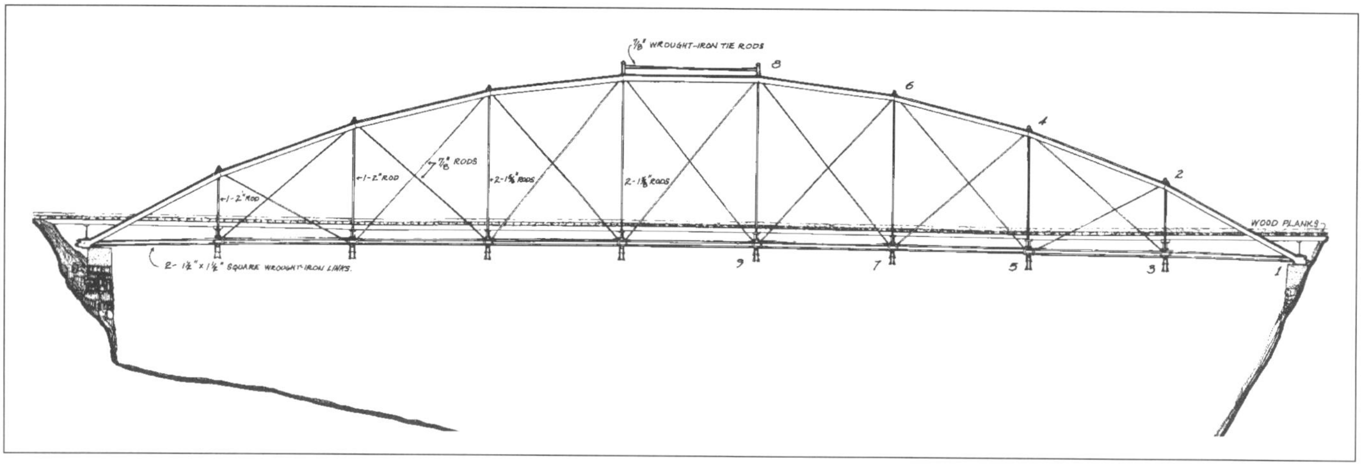

Whipple Bowstring Truss Bridge, 1867, Albany, New York.

Simon DeGraff built this cast- and wrought-iron Whipple bowstring truss bridge, based on a patent granted to bridge engineer Squire Whipple in 1841, at an unknown location in 1867. The bridge later was moved to its current location. As one historian has observed, "one of the happiest features of the bolted and pinned form of bridge construction in use before riveting became common about 1900, was not only the speed of erection, but the ease with which a span could be knocked down, moved in small pieces, and as easily and quickly re-erected on a new site."[15]

Gothic Arch Bridge (Central Park Bridge No. 28), 1864, New York City, New York.

English émigrés Calvert Vaux and Jacob Wrey Mould designed a group of spectacular cast-iron bridges for New York City's Central Park to accompany Vaux's and Frederick Law Olmsted's landscape designs for the park. The iron castings were produced by the J.B. and W.W. Cornell Ironworks of New York.

[BELOW] *Eads Bridge, 1869-74, St. Louis, Missouri.*

[RIGHT] *Railroad deck.*

The Eads Bridge is one of the most important structures of the nineteenth century. Designed by James Buchanan Eads, an extraordinarily talented man who had never designed a bridge before, the Eads Bridge marked the first use in America of steel for significant structural purposes, the first use of pneumatic (pressurized) caissons for building bridge foundations, and the innovative adoption of the cantilevering method for erecting long-span bridge arches without falsework.[16]

[ABOVE] *Poughkeepsie Bridge, 1888, Poughkeepsie, New York.*

[RIGHT] *Northampton Lattice Truss Bridge, 1887, Boston and Maine Railroad, Northampton, Massachusetts.*

New developments in metal fabrication, materials, and structural analysis enabled railroad engineers to build larger and longer bridges as the nineteenth century progressed. The Northampton Bridge was an early example of an all-riveted lattice truss, which produced a stiffer bridge than the traditional pin-connected bridge, but also one more complicated in terms of structural analysis. The Poughkeepsie Bridge, 6,767-feet long, was the longest steel structure in existence when erected in 1888 for the Central New England Railroad. This enormous long-span cantilever bridge was built to connect the coal fields of Pennsylvania with the factories of New England.[17]

Wheeling Suspension Bridge, 1849, Wheeling, West Virginia.

One of the pioneers of American suspension bridge building was Charles Ellet, Jr. The Wheeling Suspension Bridge, constructed in 1849, was his most significant contribution. The bridge was the first crossing of the Ohio River, and at more than 1,000-feet long, the longest suspension bridge in the world when completed. The bridge was heavily damaged in a storm in 1854 and has been repaired and strengthened several times. The cable stays radiating from the tops of the towers were added for stiffening purposes in the 1870s by Washington Roebling, son of Ellet's competitor, John Roebling.

Smithfield Street Bridge, 1883, 1891, Pittsburgh, Pennsylvania.

The Smithfield Street Bridge, built in Pittsburgh, Pennsylvania, in 1883 by Gustav Lindenthal, is significant because of its early use of steel and because of its interesting form. Lindenthal's bridge, consisting originally of two 360-foot steel spans, was one of the first steel truss bridges in America. A third line of trussing was added in 1891 to accommodate a streetcar line. The form of the bridge is a lenticular truss, which Lindenthal referred to as a "Pauli" truss, named after a German engineer who used the form in 1857. Lindenthal chose this particular truss for both its strength and aesthetic properties.[18]

References

1. David E. Nye, *American Technological Sublime* (Cambridge, MA: MIT Press, 1994) p. 57.

2. Carroll Pursell, *The Machine in America: A Social History of Technology* (Baltimore: Johns Hopkins University Press, 1995) p. 68.

3. Howard Mumford Jones, *The Age of Energy: Varieties of American Experience, 1865–1915* (New York: Viking, 1970) p. 152.

4. Michael Chevalier, *Society, Manners, and Politics in the United States* (1836), as quoted in Nye, *American Technological Sublime*, p. 46.

5. George Rogers Taylor, *The Transportation Revolution, 1815–1860* (New York: Armonk, 1977 [originally published in 1951]) p. 443.

6. J. C. Beltrami, *A Pilgrimage in America* (Chicago: Quadrangle Books, 1962 [originally published in 1828]) p. 65. From Letter XII, dated April 20, 1823.

7. Dan Clement, "Morris Canal," U.S. Department of the Interior, Historic American Engineering Record (HAER), No. NJ-29, Prints and Photographs Division, Library of Congress, Washington, D.C., (1983) p. 5; "Morris Canal, Inclined Plane 9 West," U.S. Department of the Interior, Historic American Engineering Record (HAER), No. NJ-29D, Prints and Photographs Division, Library of Congress, Washington, D.C., 1983, p. 1.

8. *The First Railroad in America* (Boston: Granite Railway Co., 1926), pp. 13–14.

9. Kevin Murphy, "Louisville & Nashville Railroad: Union Station Trainshed," U.S. Department of the Interior, Historic American Engineering Record (HAER), No. AL-1, Prints and Photographs Division, Library of Congress, Washington, D.C., (1984) p. 2.

10. John Ruskin, *The Stones of Venice* (1851), as quoted in Carroll L. V. Meeks, *The Railroad Station: An Architectural History* (New Haven, CT: Yale University Press, 1956), pp. 161–163.

11. Daniel L. Schodek, *Landmarks in American Civil Engineering,* (Cambridge, MA: MIT Press, 1987) pp. 165–168.

12. Spiro Kostof, *America by Design* (New York: Oxford University Press, 1987) p. 150.

13. Emory L. Kemp and Richard K. Anderson, Jr., "The Reading-Halls Station Bridge," U.S. Department of the Interior, Historic American Engineering Record (HAER), No. PA-55, Prints and Photographs Division, Library of Congress, Washington, D.C., (undated) pp. 16–17.

14. Carl W. Condit, *American Building Art: The Nineteenth Century* (New York: Oxford University Press, 1960) p. 229; Schodek, *Landmarks in American Civil Engineering,* pp. 81–83.

15. Richard S. Allen, "Whipple Cast and Wrought Iron Bowstring Truss Bridge," (HAER report NY-4), in Robert M. Vogel, ed., *A Report of the Mohawk-Hudson Area Survey* (Washington, D.C.: Smithsonian Institution Press, 1973) p. 141.

16. Condit. *American Building Art*, pp. 185–190.

17. Eric DeLony, *Landmark American Bridges* (New York: ASCE Press, 1993) p. 94; Condit, *American Building Art,* p. 158.

18. Victor Darnell, "Lenticular Bridges from East Berlin, Connecticut," *IA, Journal of the Society of Industrial Archeology* 5 (1979):19; Schodek, *Landmarks in American Civil Engineering,* p. 130.

EPILOGUE

Ruins and Remembrance

Pervasive large-scale de-industrialization began in the United States in the second half of the twentieth century, but every period of American history has witnessed mill and factory closings, the abandonment of this or that industry, and nostalgia for a lost past. Henry Longfellow lamented the passing of the familiar "village smithy" in 1839, and even as early as 1850, in America's industrial infancy, a deserted textile factory was mourned with sorrowful prose:

> There is no sight which conveys a deeper sensation of 'sadness lone,' than that of a factory, once jocund with the sound of an hundred voices, and the gleesome hurling of throstle and loom, standing tall, deserted looking, and silent. The once busy wheel, which gave motion to thousands of spindles, and hundreds of shuttles, stands gloomy and motionless, like a worn-out war steed. The bell that once clanged cheerily at the evening hour, no more calls out hundreds of gladsome toilers, gushing home through the factory doors, to enjoy the evening's recreation and repose.[1]

Such attitudes have not been common, however, nor have they necessarily led to the preservation of historic engineering and technological sites, but Americans have not been completely remiss in noticing their industrial history. Monuments and memorials have been erected at various prominent sites, such as that of the Granite Railway in Quincy, Massachusetts, and covered bridges, gristmills, iron furnace stacks, canals, and railroad-related sites and artifacts have long been of sentimental value to many Americans. Since the late 1950s, however, there has been a growing interest in both the history of American engineering and technology and an appreciation for the preservation of sites and artifacts associated with that history. The creation of organizations such as the American Society of Civil Engineers' Committee on the History and Heritage of Civil Engineering, the Society for Industrial Archeology, the Society for the History of Technology, and the Historic American Engineering Record reflects this interest, as does the growing number of museums, National Park Service sites, journals, and university graduate programs focusing on the history of American technology.

There can be little doubt that the engineers and builders of the nineteenth century technological world valued their handiwork and felt pride in their accomplishments. Historians have long recognized that there existed a "technology of haste" in American industrialization—speed, distance, and economy counted for more than permanence.[2] Yet many nineteenth-century engineering and technological objects, far from being anonymous structures, abound with the names of the engineers, builders, and companies responsible for their creation. Many builders would no doubt have agreed with Montgomery Meigs, the engineer who inscribed on one of the structures of his aqueduct system in Washington, D.C., "Esto Perpetua"—It Will Last Forever.[3]

Americans, whether they jumped or were pushed, found themselves in the "technological torrent" in the nineteenth century. Remnants of that torrent still exist, some in useful service, others memorialized, and many neglected, but all as complex

reminders of America's first century of industrialization, when the myriad values of our modern technological culture were first formed.

[ABOVE] *Former site of the Ensley Works of U.S. Steel, near Birmingham, Alabama.*

The Ensley steel mill and blast furnace plant once was a pioneering southern industrial site, where the first duplex steel in America—produced in a Bessemer convertor and then transferred to open-hearth furnaces—was made in 1899. The downturn in the U.S. steel industry spelled the end of Ensley, however, and all that remains are a few miscellaneous structures.

[RIGHT] *Schoharie Creek Aqueduct, 1841, Fort Hunter, New York.*

Buena Vista Furnace, 1847, near Armagh, Pennsylvania.

Even in neglect, some industrial structures refuse to go away. Stone blast furnaces, because of their often remote locations, sturdy materials, and construction, have a particularly high survival rate. The Buena Vista Furnace, near Armagh, Pennsylvania, was built in 1847, but because of a lack of transportation and the poor quality of the local iron ore, it closed by 1856. Vegetation has started to overtake the structure, and one corner is crumbling, but 142 years after it went out of blast, the furnace still clings to the landscape, an industrial ruin.

Remains of a "package freighter," c. 1890s, Shooters Island Ships Graveyard, Staten Island, New York.

From the mid-nineteenth century until the 1920s, Shooters Island supported small industrial and ship-building operations. After 1922, however, when the last of these concerns closed down, the island became a ship graveyard, one of five in the New York Harbor. By the 1980s, there were at least 180 derelict ship hulks abandoned in the shallows of Shooters Island.

Abandoned locomotives, c. 1880s, near Nome, Alaska.

These locomotives began life a continent away, on the elevated rail lines of Brooklyn, New York, in the 1880s. After their urban service, they were brought to the gold fields of Alaska and assumed a second life on the Council City and Solomon River Railroad.

Golden Gate Mill, c. 1898, near Ophir, Utah.

These two photographs vividly show how completely an industrial site can be abandoned. The top photograph shows the town of Mercur around 1910; in the upper left is the Golden Gate Mill, built in around 1898 by the DeLamar Mercur Mines Company for the reduction of gold ore. The bottom photograph shows the same view in 1971 when recorded by the Historic American Engineering Record.

[LEFT] *Memorial to Granite Railway, 1826, Quincy, Massachusetts.*

The original roadbed is on the left. The track in the center has been adapted for motorized trucks.

Many historic engineering sites have not been allowed to crumble into ruins and disappear, but have been preserved and memorialized. The Granite Railway Company erected these obelisks in 1921 by the incline of the Granite Railway, with plaques commemorating the railway. The Granite Railway was one of the first railroads in America, constructed in 1826 to transport stone from a quarry to the Battle of Bunker Hill monument site in Boston, Massachusetts. [4]

[ABOVE] *Cornerstone plaque, Engine House, Pumping Station No. 1, 1856, Louisville, Kentucky.*

[ABOVE] *Interior, Hook Windmill, 1806, East Hampton, New York.*

Sometimes memorialization is more informal. A hand-lettered sign in Hook Windmill reads, "This Mill is a Monument to Brain and Brawn."

[ABOVE] *Headgate House, Cohoes Company Power Canal System, 1866, Cohoes, New York.*

Officials of the Cohoes Company proudly stand before their new Headgate House in 1866, below a huge plaque that probably bears their names.

[RIGHT] *Nameplate, Main Street Bridge, 1896-97, Painesville, Ohio.*

The King Bridge Company of Cleveland, Ohio, constructed this Pennsylvania-type, pin-connected through truss bridge. Bridges were ideal public structures on which both politicians and builders could fasten a nameplate.

[ABOVE] *Maker's name cast into a tower post, Bidwell Bar Suspension Bridge, 1855, near Oroville, California.*

The Starbuck Iron Works in Troy, New York, cast the tower posts of this suspension bridge in California.

[LEFT] *Inscription on the Cabin John Aqueduct, 1859, Cabin John, Maryland.*

The Cabin John Aqueduct in Maryland illustrates the power and politics of inscriptions. Montgomery C. Meigs, the aqueduct's builder, proudly placed his name on the arch of the aqueduct during the course of construction. After Meigs was reassigned to another project, his replacement engineers inscribed *their* names on the aqueduct. Meigs returned to the project, however, and feeling the replacement engineers did not deserve credit for completing the structure, he had their names removed (see the recessed panels in the two blocks below Meigs's inscriptions). Meigs, of the United States Army Corps of Engineers, had also originally included the name of Jefferson Davis in another inscription on the aqueduct, as Davis had been the Secretary of War and Meigs's superior when work on the aqueduct had commenced in the early 1850s. During the Civil War, when Davis became President of the Confederacy, the Secretary of the Interior had Davis's name removed. His name was re-inscribed in 1909 after southern politicians protested. Meigs was also later criticized for not including in any inscription the name of one of his assistant engineers, Alfred Rives, who had helped to design and construct the aqueduct, but who also had joined the Confederacy just before Meigs had the builders' names inscribed on the aqueduct. Finally, Meigs also had inscribed on the aqueduct his name for the structure, "Union Arch," but the name never caught on, and this much-inscribed structure has always been identified by its location, spanning Cabin John Creek.

[RIGHT] *Inscription stone, Monocacy Aqueduct, 1833, Chesapeake and Ohio Canal, near Dickerson, Maryland.*

[BELOW RIGHT] *Keystone, Waverley Mill, 1857, Franklin Manufacturing Company, Paterson, New Jersey.*

[BELOW LEFT] *Cast-iron fender of doorway, Martinsburg Machine Shop, 1866, Baltimore and Ohio Railroad, Martinsburg, West Virginia.*

[FACING PAGE]
Entrance, Granite Hydroelectric Plant, 1896, Utah Power Company, near Holladay, Utah.

UTAH POWER COMPANY
ERECTED
1896
NO ADMITTANCE
WITHOUT PERMISSION
FROM SUPERINTENDENT
DANGER
VOLTAGE

[RIGHT] *Lexington Depot, c. 1846, Lexington and West Cambridge Railroad, Lexington, Massachusetts.*

Many industrial and engineering structures in the nineteenth century were reused by other industrial concerns, but several have found completely new uses in the twentieth century. The Lexington Depot is a rare survivor of a railroad station form that included track space beneath the station roof. The station was given a Colonial Revival makeover in the 1920s and was later converted into a bank.

[LEFT] *Eureka Bridge, c. 1872, Castalia, Iowa.*

The Eureka Bridge is a type of bowstring truss patented in 1871 by William Allen and Oliver Perry and manufactured by the firm of Allen, McEvoy & Company, machinists and general jobbers located in nearby Beloit, Wisconsin. The original location of the bridge is unknown, but it was moved in the early 1990s from near Frankville, Iowa, to the city park of Castalia, Iowa.[5]

Metacomet Mill, 1847,
Fall River, Massachusetts.

Trapped in a twentieth-century cloverleaf, the Metacomet Mill in Fall River, Massachusetts, still stands as a reminder of New England textile manufacturing in the nineteenth century. The Metacomet Mill began operations in 1847, helping to make Fall River a major textile center. By the 1930s, however, the New England textile industry was in depression as mills moved south, and buildings such as the Metacomet Mill became home to an assortment of small industries.

[ABOVE] *Station Road Bridge (in foreground), 1882, and its replacement, the Brecksville-Northfield High Level Bridge (in background), 1931, Brecksville, Ohio.*

Change, alteration, modernization, obsolescence—few engineering and industrial structures retain their place in the scheme of things for long. New materials, new techniques, new needs come along, and the old is abandoned. As we enter a new millennium, we might remember that only a fraction of the nineteenth-century engineering and industrial world still exists, silent reminders of our industrial legacy, of gear-and-girder technology, and of a world that was remade by a technological transformation.

References

1. "Our Manufactures," *Scientific American* 6 (December 7, 1850), as quoted in John F. Kasson, *Civilizing the Machine: Technology and Republican Values in America, 1776–1900* (New York: Penguin, 1976) p. 140.

2. Daniel J. Boorstin, *The Americans: The National Experience* (New York: Vintage, 1965) p. 106.

3. William T. S. Curtis, "Cabin John Bridge," *Records of the Columbia Historical Society* 2 (1899):307.

4. *The First Railroad in America* (Boston: Granite Railway Co., 1926), pp. 13–14, 21.

5. Eric DeLony, *Landmark American Bridges* (New York: ASCE Press, 1993) p. 66.

Bibliography

Allen, Richard S. "Whipple Cast and Wrought Iron Bowstring Truss Bridge 1867." In *A Report of the Mohawk-Hudson Area Survey*, edited by Robert M. Vogel, 137-150. Washington, DC: Smithsonian Institution Press, 1973.

American Society of Mechanical Engineers International History and Heritage Committee. *Landmarks in Mechanical Engineering*. West Lafayette: Purdue University Press, 1997.

Athan, Polly. "Technological Watersheds on the Waupaca River: A History of the Fisher-Fallgatter Mill," in U.S. Department of the Interior, Historic American Engineering Record (HAER), No. WI-1, "Fisher-Fallgatter Mill," 1979, Prints and Photographs Division, Library of Congress, Washington, D.C.

Bailey, Thomas A. and David M. Kennedy. *The American Pageant, Vol. II*. 8th ed. Lexington, MA: D.C. Heath, 1987.

Banham, Reyner. "The Becher Vision." In Bernd and Hilla Becher, *Water Towers,* 7-8. Cambridge, MA: MIT, 1988.

Bathe, Greville and Dorothy Bathe. *Oliver Evans: A Chronicle of Early American Engineering*. Philadelphia: Historical Society of Pennsylvania, 1935.

Beltrami, J.C. *A Pilgrimage in America*. Chicago: Quadrangle Books, 1962 [originally published 1828].

Bennett, Lola. U.S. Department of the Interior, Historic American Engineering Record (HAER), No. MA-119, "Coleman Bridge," 1990, Prints and Photographs Division, Library of Congress, Washington, D.C.

Berlin, Isaiah. *The Hedgehog and the Fox*. New York: Simon and Schuster, n.d. [reprint of 1953 edition].

Biggs, Lindy. *The Rational Factory: Architecture, Technology, and Work in America's Age of Mass Production*. Baltimore: Johns Hopkins University Press, 1996.

Boorstin, Daniel J. *The Americans: The National Experience*. New York: Vintage, 1965.

Borgstrom, G. "Food and Agriculture in the Nineteenth Century." In *Technology in Western Civilization,* Vol. I, edited by Melvin Kranzberg and Carroll W. Pursell, Jr., New York: Oxford University Press, 1967, p. 423.

Brown, Richard D. "Modernization: A Victorian Climax." In *Victorian America*, edited by Daniel Walker Howe, 29-44. Philadelphia: University of Pennsylvania, 1976.

Clement, Dan. U.S. Department of the Interior, Historic American Engineering Record (HAER), No. NJ-29, "Morris Canal," 1983, and No. NJ-29D, "Morris Canal, Inclined Plane 9 West," 1983, Prints and Photographs Division, Library of Congress, Washington, D.C.

Cochran, Thomas C. and William Miller. *The Age of Enterprise: A Social History of Industrial America*. Revised ed. New York: Harper, 1961.

The Complete Poetical Works of Henry Wadsworth Longfellow. Boston: Houghton, Mifflin, 1902.

Condit, Carl W. *American Building Art: The Nineteenth Century*. New York: Oxford, 1960.

———. *American Building*. Chicago: University of Chicago, 1968.

Cowan, Ruth Schwartz. *A Social History of American Technology*. New York: Oxford, 1997.

Curtis, William T.S. "Cabin John Bridge." *Records of the Columbia Historical Society* 2 (1899):293-307.

Darnell, Victor. "Lenticular Bridges from East Berlin, Connecticut." *IA- Journal of the Society for Industrial Archeology*, 5 (1979):19-32.

DeLony, Eric. *Landmark American Bridges*. New York: American Society of Civil Engineers, 1993.

The First Railroad in America. Boston: Granite Railway Co., 1926.

Giedion, Siegfried. *Mechanization Takes Command*.(New York: Norton, 1969 [originally published in 1948].

Gordon, Robert B. and Malone, Patrick M., *The Texture of Industry: An Archaeological View of the Industrialization of North America* (New York: Oxford University Press, 1994) pp. 312–315

Gross, Laurence F. U.S. Department of the Interior, Historic American Engineering Record (HAER), No. MO-1, "Watkins (Woolen) Mill," 1978, Prints and Photographs Division, Library of Congress, Washington, D.C.

———. "The Importance of Research Outside the Library: Watkins Mill, A Case Study." *IA, Journal of the Society for Industrial Archeology* 7 (1981):15-26.

Hindle, Brooke. "How Much Is a Piece of the True Cross Worth?" In *Material Culture and the Study of American Life*, edited by Ian M.G.Quimby, 5-20. New York: Norton, 1978.

———. "The Exhilaration of Early American Technology." In *Early American Technology: Making and Doing Things from the Colonial Era to 1850*, edited by Judith A. McGaw, 40-67. Chapel Hill: University of North Carolina, 1994.

——— and Steven Lubar. *Engines of Change: The American Industrial Revolution, 1790-1860*. Washington: Smithsonian Institution, 1986.

Hofstadter, Richard William Miller, and Daniel Aaron. *The United States*. Part 2. Englewood Cliffs, NJ: Prentice-Hall, 1972.

Hounshell, David. *From the American System to Mass Production, 1800-1932*. Baltimore: Johns Hopkins, 1984.

Hughes, Thomas P. *American Genesis: A Century of Invention and Technological Enthusiasm, 1870-1970*. New York: Viking, 1989.

———, ed. *Changing Attitudes Toward American Technology*. New York: Harper and Row, 1975.

Hunter, Louis C. *A History of Industrial Power in the United States, 1780-1930, Vol. I: Waterpower in the Century of the Steam Engine*. Charlottesville, VA: University Press of Virginia, 1979.

———. *A History of Industrial Power in the United States, Vol. II: Steam Power.* Charlottesville, VA: University Press of Virginia, 1985.

——— and Lynwood Bryant. *A History of Industrial Power in the United States, Vol. III: The Transmission of Power.* Cambridge, MA: MIT, 1991.

Hurley, F. Jack, ed. *Industry and the Photographic Image*. New York: Dover, 1980.

Jones, Howard Mumford. *The Age of Energy: Varieties of American Experience, 1865-1915*. New York: Viking, 1970.

Kasson, John F. *Civilizing the Machine: Technology and Republican Values in America, 1776-1900*. New York: Penguin, 1976.

Kemp, Emory L. and Richard K. Anderson, Jr. U.S. Department of the Interior, Historic American Engineering Record (HAER), No. PA-55, "The Reading-Halls Station Bridge," (undated), Prints and Photographs Division, Library of Congress, Washington, D.C.

Kostof, Spiro. *America by Design*. New York: Oxford, 1987.

Kranzberg, Melvin and Carroll W. Pursell, Jr., eds. *Technology in Western Civilization, Vol. I*. New York: Oxford University Press, 1967.

Larson, Andrew Karl. *I Was Called to Dixie*. Salt Lake City: Deseret News Press, 1961.

Lewis, W. David. *Sloss Furnaces and the Rise of the Birmingham District: An Industrial Epic*. Tuscaloosa, AL: University of Alabama, 1994.

Licht, Walter. *Industrializing America: The Nineteenth Century.* Baltimore: Johns Hopkins, 1995.

Lunden, Walter A. "The Rotary Jail, or Human Squirrel Cage." *Journal of the Society of Architectural Historians*, XVIII (December 1959):149-157.

Malone, Patrick M. "Little Kinks and Devices at Springfield Armory, 1892-1918." *IA, Journal of the Society for Industrial Archeology* 14 (1988):59-76.

Marcus, Alan I. and Howard P. Segal. *Technology in America: A Brief History.* San Diego: Harcourt Brace Jovanovich, 1989.

Marx, Leo. "The Idea of 'Technology' and Postmodern Pessimism." In *Technology, Pessimism, and Postmodernism*, edited by Yaron Ezrahi, Everett Mendelsohn, and Howard Segal, Dordrecht: Kluwer, 1994.

McGrain, John W. "'Good Bye Old Burr': The Roller Mill Revolution in Maryland, 1882." *Maryland Historical Magazine*, 77 (Summer, 1982):154-171.

Meeks, Carroll L. V. *The Railroad Station: An Architectural History.* New Haven: Yale, 1956.

Miller, Perry. "The Responsibility of Mind in a Civilization of Machines." In *Changing Attitudes Toward American Technology*, edited by Thomas P. Hughes, 63-83. New York: Harper and Row, 1975.

Morison, Elting. *From Know-How to Nowhere: The Development of American Technology.* New York, Basic, 1974.

Mumford, Lewis. *Technics and Civilization*. New York: Harcourt, Brace, 1934.

Murphy, Kevin. U.S. Department of the Interior, Historic American Engineering Record (HAER), No. AL-1, "Louisville & Nashville Railroad: Union Station Trainshed," 1984, Prints and Photographs Division, Library of Congress, Washington, D.C.

Nye, David E. *American Technological Sublime*. Cambridge, MA: MIT, 1994.

Nye, Russell Blaine. *Society and Culture in America, 1830-1860*. New York: Harper, 1974.

O'Connor, Richard. U.S. Department of the Interior, Historic American Engineering Record (HAER), No. LA-11, "Magnolia Plantation Cotton Gins and Presses," 1997, Prints and Photographs Division, Library of Congress, Washington, D.C.

Oestreicher, Richard. "Images of Ideology and the Ideology of Imagery." *Labour/Le Travail* 15 (Spring 1985):179-183.

Orvell, Miles. *The Real Thing: Imitation and Authenticity in American Culture, 1880-1940*. Chapel Hill, NC: University of North Carolina Press, 1989.

Penn, Theodore Z. "The Development of the Leather Belt Main Drive." *IA, Journal of the Society for Industrial Archeology* 7 (1981):1-14.

Pursell, Carroll. *Early Stationary Steam Engines in America*. Washington: Smithsonian Institution, 1969.

———. *The Machine in America: A Social History of Technology* (Baltimore: Johns Hopkins University Press, 1995.

Reynolds, David S. *Walt Whitman's America*. New York: Alfred A. Knopf, 1995.

Riggs, Connie. "The Human Squirrel Cage." *Indiana History Bulletin* 53 (October 1975):123-126.

Rose, Kenneth D. U.S. Department of the Interior, Historic American Engineering Record (HAER), No. WA-19, "Goat Trail [Mining Road] at Devil's Corner," 1987, Prints and Photographs Division, Library of Congress, Washington, D.C.

Rumm, John. U.S. Department of the Interior, Historic American Engineering Record (HAER), No. VI-7, "The Development of the Sugar Cane Industry in the Virgin Islands," 1977, Prints and Photographs Division, Library of Congress, Washington, D.C.

Sande, Theodore Anton. *Industrial Archeology - A New Look at the American Heritage*. New York: Penguin, 1976.

Schodek, Daniel L. *Landmarks in American Civil Engineering*. Cambridge, MA: MIT, 1987.

Schurr, Samuel H. *Energy in the American Economy, 1850-1955*. Baltimore: Johns Hopkins, 1960.

Seely, Bruce E. "Blast Furnace Technology in the Mid-19th Century: A Case Study of the Adirondack Iron and Steel Company." *IA, Journal of the Society for Industrial Archeology* 7 (1981):27-54.

Shoup, Laurence H. "A Historiographic Essay and Annotated Bibliography of the Folsom Hydroelectric Power Complex, Sacramento County, California." 1993 [Unpublished paper; copy in the office of the Historic American Engineering Record].

Stevens, Edward W. Jr., *The Grammar of the Machine: Technical Literacy and Early Industrial Expansion in the United States* (New Haven, CT: Yale University Press, 1995).

Stott, Peter. *A Guide to the Industrial Archeology of Boston Proper*. Cambridge, MA: MIT, 1984.

Taylor, George Rogers. *The Transportation Revolution, 1815-1860*. Armonk, New York: 1977 [originally published in 1951].

Tichi, Cecilia. *Shifting Gears: Technology, Literature and Culture in Modernist America*. Chapel Hill, NC: University of North Carolina, 1987.

Trachtenberg, Alan. *The Incorporation of America: Culture and Society in the Gilded Age*. New York: Hill and Wang, 1982.

United States Department of Commerce, Bureau of the Census. *Historical Statistics of the United States, Colonial Times to 1970*. Washington: GPO, 1975.

Vlach, John Michael. *Back of the Big House - The Architecture of Plantation Slavery.* Chapel Hill: University of North Carolina Press, 1993.

Vogel, Robert. "The Burden Water-Wheel." *Occasional Publications* (Society for Industrial Archeology), No.2, April 1973.

Wallace, Kim E. *Brickyard Towns: A History of Refractories Industry Communities in South-Central Pennsylvania*. Washington: National Park Service, 1993.

Watson, Aldren A. *The Village Blacksmith*. New York: Thomas Y. Crowell, 1968.

Wiebe, Robert H. *The Search for Order, 1877-1920*. New York: Hill and Wang, 1967.

Image Credits

Credits include HABS-HAER catalog number

Abbreviations: (t) – top; (b) – below; (r) – right; (l) – left; (a) – above; (m) – middle; (fl) – far left; (tl) – top left; (tr) – top right; (bl) – bottom left; (br) – bottom right; (in) – inset

Cover	William E. Barrett, 1970, MD-1
Frontispiece	William E. Barrett, 1973, WV-5

PROLOGUE

Page	
10	Jet Lowe, 1982, NY-18
11(r)	Jet Lowe, 1985, VI-6
11(l)	J. Carl Burke, 1970, NY-61
12	Jet Lowe,1982, MA-16
13(r)	Jack Boucher, 1976, MA-8A
13(l)	Courtesy of the National Museum of American History, Smithsonian Institution, MD-29
14	Jack Boucher, 1978, PA-51
15(l)	Jet Lowe, 1985, NY-138
15(b)	Jet Lowe, 1988, NY-138
16(r)	Jet Lowe, 1986, NJ-70C
16(b)	Jet Lowe, 1986, NJ-70A

CHAPTER I

Page	
19	Jet Lowe, 1978, NY-105
20	Jack Boucher, 1987, SC-13
21(a)(l)	Ned Goode, 1961, HABS PA-116
22(a)(r)	Charles N. Bayless, 1977, HABS SC-476-B
23(l)(b)	Jack Boucher, 1987, SC-13
24(a)	Christopher H. Marston and Thomas Behrens, 1997, LA-11
24(r)	John Nicely, 1996, LA-11
25(r)(b)	John Nicely, 1996, LA-11
26	Brian Grogan, 1991, CA-119
27	John R. Bowie, 1986, PA-104

28(t)(b)	William E. Barrett, 1973, WV-5
29(a)	Courtesy of Gwendolyn G. de Clairville, Huntington, NY, NY-106
29(l)	Kathleen Hoeft, 1975, NY-106
30	Jet Lowe, 1979, WI-1
31	Susan Keil, Ellen Stoner, and Gil Witte, 1987, IL-25
32(a)	Jet Lowe, 1985, NY-240
32(l)	Patricia Reese and Craig Strong, 1990, NY-240
33(a)	Patricia Reese and Craig Strong, 1990, NY-240
33(r)	Jet Lowe, 1990, NY-240
34	Gary C. Long, 1976, NY-67
35(a)(b)	Jet Lowe, 1978, NY-143
35(r)	Kathleen S. Hoeft and Chalmers G. Long, Jr. 1976, NY-67
36(a)	Kathleen S. Hoeft, 1976, NY-134
36(fl)	Jet Lowe, 1978, NY-67
36(l)	W. Richard Ansteth, 1975, NY-105
37	Jack Boucher, 1974, HABS WY-42
38	Martine Dion, 1992, CA-128
39(l)	Paul Anderson, 1987, MT-36C
39(b)	Courtesy of the Kenneth March Collection, Michigan Technological University, MI-2
40	Jet Lowe, 1979, MT-36B
41(t)	Jet Lowe, 1980, MT-36B
41(l)	Jet Lowe, 1993, AL-80B
42(a)	Jet Lowe, 1990, LA-9
42(l)	Mercedes Garzon Maceda and Albert Debnam, 1989-90, LA-9
43(r)(b)	Jet Lowe, 1990, LA-9
44(r)(b)	William E. Barrett, 1973, WV-9
45	Scott W. Daley, 1997, PA-441
46	Jack Boucher, 1969, HABS WI-249
47(r)	Larry Hermsen, 1969, HABS WI-249
47(l)	Jack Boucher, 1969, HABS WI-249
48	Jack Boucher, 1975, KY-9
49(tr)	Jack Boucher, 1986, HABS MO-1671
49(tl)	Peter Stott, MA-25
49(br)	Jack Boucher, 1986, HABS MO-1832A
49(bl)	Photographer unknown, c.1964, HABS IL-1041
50	Greg Anderson, 1987, MN-19
51	Jack Boucher, 1978, NY-132
52(a)	Jack Boucher, 1969,NY-2
52 (bl)	Jack Boucher, 1969, NY-2
52 (br)	Eric Delony, 1969, NY-2
53(a)(ar)(r)	Gary Sampson, 1982, NH-7
54 (l)(b)	Jack Boucher, 1971, VA-14

CHAPTER 2

Page	
57	Jet Lowe, 1978, MO-1
58	Original photograph courtesy of the Manchester Historic Association, Manchester, NH, HABS NH-109
59(a)	Joseph Elliott, 1991, RI-1
59(r)	Robert C. Griebner and Todd A. Croteau, 1991-1992, RI-1
60(a)	R. Randolph Langenbach, 1967, HABS RI-338
60(r)	Jet Lowe, 1978, NH-2A
61	R. Randolph Langenbach, 1967, HABS NH-109
62(r)(b)	Jet Lowe, 1998, SC-27
63(a)(l)	Jack Boucher, 1974, IN-1
64(l)(b)	Jet Lowe, 1978, MO-1
65(a)(r)	Jack Boucher, 1969, NY-5
66(a)(r)	David Sharpe, 1977, GA-19

67	Jet Lowe, 1991, PA-265
68	John R. Bowie and Dianne M. Pohlsander, 1989, MD-76
69	Jet Lowe, 1991, PA-257
70	Jet Lowe, 1978, NY-123
71(t)	John R. Bowie, 1978, NY-123
71(b)	Barry A. Richards, 1978, NY-123
72(l)	Alex Bush, 1936, HABS AL-276
72(b)	Jet Lowe, 1993, AL-122
73	Jack Boucher, 1974, AL-3
74	Jet Lowe, 1989, PA-115A
75	Christopher H. Marston, 1990, PA-115
76	Robert D. Welch, 1981, UT-25A
77(l)	Nathaniel R. Ewan, 1939, HABS NJ-560
77(b)	Jack Boucher, 1972, UT-7
78	Jet Lowe, 1991, PA-260
79	Robington Emmanuel, 1993, PA-283
80	David Ames, 1982, HABS DE-2
81	Jet Lowe, 1990, MN-46
82(t)	Photocopy by William E. Barrett, 1973 from original Franklin P. Gruber, PA-14
82(b)	William E. Barrett, 1974, PA-14
83(r)(b)	William E. Barrett, 1974, PA-14
84(r)	Frederick R. Love, 1973, WV-6
84(b)	Jet Lowe, 1979, PA-156
85	Dewey G. Mears, 1968, HABS TX-3173
86	Original photograph courtesy of the Aurora Historical Society, Aurora, IL, IL-8
87(r)(bl)	Jet Lowe, 1986, MA-16
87(br)	Jack Boucher, 1969, HABS RI-303
88(r)	Original photograph courtesy of Mrs. L.J. Fargo, Lexington, OH, OH-24
88(b)	Original photograph courtesy of Roots-Connersville Blower Company, Connersville, IN, IN-3
89(a)	Original photograph courtesy of Mrs. Robert Cummins, Cannelton, IN, IN-1C
89(l)	Jack Boucher, 1974, IN-1C
90	Jet Lowe, 1990, LA-9
91(l)	Original photograph courtesy of Rensselear Polytechnic Institute Library, Troy, NY, NY-7B
91(r)	John R. Bowie, 1986-87, PA-104
92(r)	Photocopy, Jack Boucher, 1969, NY-5; original from Evan Leigh, *Modern Cotton Spinning* (Vol. 1), Manchester (1873), plate XXI
92(l)	Fred Gjessing, 1976, PR-1A
93	Robert Levy, 1977, MD-39
94(r)	Pierce Bounds, 1983, PA-61
94(b)	Marian Dombroski, 1983, PA-61
95	Jack Boucher, 1976, MA-8A
96(r)	Jack Boucher, 1986, PA-127A
96(b)	Jack Boucher, 1994, PA-127A
97(a)	Pat Reese, 1989, PA-127A
97(l)	Jet Lowe, 1982, MA-24
98(a)(l)	Brian Grogan, 1993, CA-66
99	Brian Grogan, 1993, CA-66
100	David Sharpe, 1974, NY-70

CHAPTER 3

Page	
103	Joseph Elliott, 1995, IA-59
103(in)	Joseph Elliott, 1991, RI-1
104(a)	Jet Lowe, 1982, MA-16

104(r) Michael Spozarsky, 1979, NJ-27C
105(a) Robert J. McNair and Toni K. Ristau, 1971, UT-1
105(r) Jack Boucher, 1971, UT-1
106(a) Lloyd Grotjan, 1987, MO-31
106(l) Jack Boucher, 1976, GA-3
107(a) Charles Parrott, 1969, NY-1
107(r)(b) Jack Boucher, 1969, NY-1
108(t)(br) Jack Boucher, 1972, NY-20
108(bl) Willy Sclarsic, 1971, NY-20
109(l) Jack Boucher, 1972, NY-20
109(b) Dale Flick, 1971, NY-20
110 Stephen Zane, 1975, NY-104
111(r) William E. Barrett, 1970, WV-1A
111(b) Jeffrey Jenkins, 1970, WV-1A
112(r)(b) William E. Barrett, 1971, MD-6
113(a) Charles Parrott, 1970, MD-2B
113(r) William E. Barrett, 1971, MD-2B
114(al) William E. Barrett, 1974, WV-3
114(ar) Hedrich-Blessing Photographers, 1976, IL-6
115(l)(b) Jack Boucher, 1976, GA-2
116(r)(b) Jack Boucher, 1974, IN-17
117(l) Jet Lowe, 1987, MN-25
117(a) Jet Lowe, 1984, CA-33
118(a) Joseph Elliott, 1995, IA-15
118(r) Erick McEvoy, 1995, IA-15
119(r)(b) Allen Lubow, 1975, NY-82
120(tr) Jack Boucher, 1988, MA-86
120(br) N.E. Baldwin, 1938, HABS NY-3250
121(t) Jet Lowe, 1994, NY-277
121(m) Martin Stupich, 1990, MA-119
121(b) Courtesy of Weatherford Public Library Archives, Weatherford, Texas, TX-36
122(r)(b) Jack Boucher, 1969, VT-1A
123(a)(l) Jet Lowe, 1987, WA-19
124(al)(ar) Jack Boucher, 1971, UT-17
125 Jet Lowe, 1978, MO-1
126(l) Jack Boucher, 1986, SC-14
126(r) Randall Page, 1977, NC-7
127(lt) Jack Boucher, 1975, KY-9
127(lb) Jack Boucher, 1971, VA-14
127(a) Jack Boucher, 1969, HABS RI-299
128(a) Martin Stupich, 1987, MA-76
128(r) Jet Lowe, 1984, NY-194

CHAPTER 4

Page
131 Courtesy of Wyoming State Museum, Cheyenne, WY, NE-6
132 Jet Lowe, 1984, ME-3
133(l) Photocopy by Jack Boucher, 1971, NJ-29 of original photograph courtesy of Canal Society of New Jersey
133(r) Jack Boucher, 1969, NY-14
134(tr) Jack Boucher, 1960, HABS MD-810
134(br) Jack Boucher, 1959, HABS MD-19
135(l) Scott Barber, Brian D. Bartolomew, Elizabeth F. Knowlan, Anne Guerette, 1988, and Dana Lockett, 1991, PA-1
135(b) David Plowden, 1969-70, PA-1
136(a)(l) J.F. Brooks, 1967, HABS FL-189
137 Jay L. Baker, 1987, MD-54

138(l) Lee Ann Jackson and Isabel C. Yang, 1988, RI-27
138(b) Martin Stupich, 1988, RI-27
139 Jet Lowe, 1985, WA-14
140(a) Robbyn L. Jackson, 1990, CA-54
140(r) Jet Lowe, 1988, CA-54
141 Jet Lowe, 1987, MN-35
142(r)(b) Arthur C. Haskell, 1934, HABS MA-150
143 Jack Boucher, 1974, AL-1
144(a) Jack Boucher, 1971, NY-50
144(l) Jack Boucher, 1971, NY-56
145 Jack Boucher, 1961, MD-14
146(r)(b) Jack Boucher, 1976, GA-1
147(a)(r) William E. Barrett, 1983, MA-58
148(b)(r) Jack Boucher, 1973-74, NY-68
149 Jack Boucher, 1971, IL-1
150 Jack Boucher, 1971, VA-2
151(a) Jack Boucher, 1986, PA-127B
151(b) Jack Boucher, 1986, PA-127
152(a) Jack Boucher, 1986, PA-127
152(rt) Jack Boucher, 1994, PA-127D
152(rb) Jack Boucher, 1986, PA-127J
153 Robert Dixon, Isabel Yang, and Thomas Behrens, 1994, PA-127D
154 Jack Boucher, 1971, VA-1
155(t) William E. Barrett, 1982, MD-41
155(b) Christopher H. Marston, 1992, PA-72
156 Arthur C. Haskell, 1934, HABS MA-101
157(a) William E. Barrett, 1970-71, MD-9
157(r) Peter Stott, 1983, MA-27
158(a)(r) Jet Lowe, 1984, PA-55
159 Jack Boucher, 1971, PA-6
160(t) David Bouse, 1969, NY-4
160(b) Jack Boucher, 1969, NY-4
161 Jet Lowe, 1984, NY-194
162(b)(r) Jet Lowe, 1983, MO-12
163(a) Jack Boucher, 1978, NY-131
163(r) Jet Lowe, 1984, MA-55
164(t) William E. Barrett, 1976, WV-2
164(b) Jack Boucher, 1974, PA-2

EPILOGUE

Page

167(a) Jet Lowe, 1994, AL-52
167(r) Jack Boucher, 1969, NY-6
168(a) Jet Lowe, 1991, PA-265
168(b) Charles Wisniewski, 1985, NY-162C
169 Jet Lowe, 1981, AK-16
170(a) Courtesy Utah Power and Light, UT-10
170(b) Jack Boucher, 1971, UT-10
171(l) Arthur C. Haskell, 1934, HABS MA-150
171(bl) William G. Johnson, 1987, KY-9
171(br) W. Richard Ansteth, 1975, NY-105
172(a) Original photograph courtesy of Niagara Mohawk Power Corporation, Albany, NY, NY-8
172(r) Wayne B. Ford, 1984, OH-34
173(a) Jet Lowe, 1984, CA-34
173(l) Russell Jones, 1959, HABS MD-180
174(r) Jack Boucher, 1959, HABS MD-19
174(br) Jack Boucher, 1973, NJ-7

174(bl)	William E. Barrett, 1970, WV-1C
175	Jack Boucher, 1971, UT-4
176(r)	Peter Stott, 1983, MA-21
176(l)	Joseph Elliott, 1995, IA-10
177	Jack Boucher, 1968, HABS MA-983
178	Jet Lowe, 1985, OH-67

Index